SpringerBriefs in Geography

For further volumes:
http://www.springer.com/series/10050

Christopher J. Parker

The Fundamentals of Human Factors Design for Volunteered Geographic Information

Springer

Christopher J. Parker
Loughborough Design School
Loughborough University
Loughborough, Leicestershire
UK

ISSN 2211-4165 ISSN 2211-4173 (electronic)
ISBN 978-3-319-03502-4 ISBN 978-3-319-03503-1 (eBook)
DOI 10.1007/978-3-319-03503-1
Springer Cham Heidelberg New York Dordrecht London

Library of Congress Control Number: 2013953690

© The Author(s) 2014

This work is subject to copyright. All rights are reserved by the Publisher, whether the whole or part of the material is concerned, specifically the rights of translation, reprinting, reuse of illustrations, recitation, broadcasting, reproduction on microfilms or in any other physical way, and transmission or information storage and retrieval, electronic adaptation, computer software, or by similar or dissimilar methodology now known or hereafter developed. Exempted from this legal reservation are brief excerpts in connection with reviews or scholarly analysis or material supplied specifically for the purpose of being entered and executed on a computer system, for exclusive use by the purchaser of the work. Duplication of this publication or parts thereof is permitted only under the provisions of the Copyright Law of the Publisher's location, in its current version, and permission for use must always be obtained from Springer. Permissions for use may be obtained through RightsLink at the Copyright Clearance Center. Violations are liable to prosecution under the respective Copyright Law.
The use of general descriptive names, registered names, trademarks, service marks, etc. in this publication does not imply, even in the absence of a specific statement, that such names are exempt from the relevant protective laws and regulations and therefore free for general use.
While the advice and information in this book are believed to be true and accurate at the date of publication, neither the authors nor the editors nor the publisher can accept any legal responsibility for any errors or omissions that may be made. The publisher makes no warranty, express or implied, with respect to the material contained herein.

Springer is part of Springer Science+Business Media (www.springer.com)

*This book is dedicated to Susan Parker,
who sadly passed away during the
research period. Through her life she
was inspirational, and without whom
this work would never have been possible.*

Susan Joy Parker
(11/09/1954–27/04/2011)

Contents

1	**Introduction**	1
	1.1 The Rise of Volunteered Information	1
	1.2 The Fundamental Issues	2
	1.2.1 Neogeography, Volunteers and Users	2
	1.2.2 Users of Volunteered Information	4
	1.2.3 Data Richness of Volunteered Information	5
	1.2.4 Trust in Volunteered Information	6
	1.3 General Aim of Book	7
	References	7
2	**A Framework of Neogeography**	11
	2.1 Introduction	11
	2.2 Background Literature	12
	2.2.1 The Nature of Neogeography	12
	2.2.2 Issues with Current Taxonomies	14
	2.3 A Framework of Neogeography	15
	2.3.1 A Terminology of Neogeography	15
	2.3.2 A Framework for Neogeography	16
	2.4 Discussion	18
	2.5 Conclusion	20
	References	20
3	**Scoping Study: User Perceptions of VGI in Neogeography**	23
	3.1 Introduction	23
	3.2 Aims	23
	3.3 Study Rationale	24
	3.3.1 Selection of VGI Platforms	24
	3.3.2 Investigation Overview	25
	3.4 Part A: Participatory Observation	25
	3.4.1 Methods	25
	3.4.2 Results and Analysis	27

3.5	Part B: Interviews		29
	3.5.1	Methods	29
	3.5.2	Results and Analysis	32
	3.5.3	Multidimensional Value Dimensions	34
	3.5.4	General User Perspectives	38
3.6	Discussion		40
	3.6.1	User Value Dimensions	40
	3.6.2	Spatial-Data Infrastructure (SDI) Relationships	42
3.7	Conclusions		42
	3.7.1	Relating to the Project Aims	42
	3.7.2	Relating to the Research Questions	45
References			45

4 Study Two: Understanding Design with VGI Using an Information Relevance Framework ... 49

4.1	Introduction		49
4.2	Aims		50
4.3	Study Rationale		50
	4.3.1	Selection of Study Community	50
	4.3.2	Investigation Overview	51
4.4	Study Two A: Participatory Observation		51
	4.4.1	Methods	51
	4.4.2	Results and Analysis	53
4.5	Study Two B: Focus Groups		54
	4.5.1	Methods	54
4.6	Discussion		65
	4.6.1	Impact of Information Depth and Scope in Understanding the Outdoor Environment	65
	4.6.2	Influence of Information Currency	70
	4.6.3	Importance of Real Time Information	71
	4.6.4	Importance of Information Access	72
	4.6.5	Importance of Trust in Information	72
	4.6.6	Volunteer Reporting of Activity Experiences	74
4.7	Conclusions		74
References			76

5 Data Generation: VGI and PGI Data Sets ... 79

5.1	Introduction		79
5.2	Research Aims		80
5.3	Study Rational		80
	5.3.1	Selection of a Study Community	80
	5.3.2	Selection of a Geographic Location for Research	81
	5.3.3	Selection of Travel Routes	81
	5.3.4	Selection of the Mashup Base Map	81

	5.4	Investigation Overview	82
	5.5	Part A: VGI Data	82
		5.5.1 Methods	82
		5.5.2 Results and Analysis	85
	5.6	Part B: PGI Data	85
		5.6.1 Methods	85
		5.6.2 Results and Analysis	86
	5.7	Mashups	87
	5.8	Discussion	89
		5.8.1 Content of Collected Data	89
		5.8.2 Success of Data Collection	91
	References		92
6	**Study Three: Assessing the Impact of VGI**		95
	6.1	Introduction	95
	6.2	Research Aims	96
	6.3	Study Rational	96
	6.4	Methodology	96
		6.4.1 Overview	96
		6.4.2 Experimental Variables	97
		6.4.3 Experimental Design	98
		6.4.4 Design of the User Judgement Survey	98
		6.4.5 Design of the Website	100
		6.4.6 Participant Sampling	104
		6.4.7 Procedure	106
		6.4.8 Statistical Analysis	107
	6.5	Hypotheses	109
	6.6	Results and Analysis	109
		6.6.1 Two-Way MANOVA	109
		6.6.2 Sample Size Estimation	111
		6.6.3 Testing of Independent Variables	112
	6.7	Discussion	113
		6.7.1 Influence of VGI on Quality and Authority	113
		6.7.2 Influence of VGI on Currency	115
		6.7.3 Sample Size Estimation	116
	6.8	Conclusions	117
	References		118
7	**Conclusion**		121
	7.1	The Nature of VGI as Distinct from PGI	121
		7.1.1 Data Content, Use and Contribution	121
		7.1.2 Information Judgements	124
	7.2	An Appraisal of the Framework of VGI	125

	7.3	Unique Influences of VGI on the User....................	127
	7.4	Limitations Of VGI from a Human Factors Perspective.......	128
	7.5	Design Recommendations for Utilising VGI................	130
	References..	131

Further Reading.. 135

Index... 137

Chapter 1
Introduction

1.1 The Rise of Volunteered Information

The concept of non-professionals having a profound impact on the nature and language of Geographic Information (GI) is not a new phenomenon. In 1507 cartographer Martin Waldseemüller drew an outline of a continent and labelled it *America*. While such an action by skilled cartographers is not in itself remarkable, Waldseemüller was particularly influenced by the Soderini Letter, the work of the *amateur* Amerigo Vespucci, a claimant for the continents' discovery (Goodchild 2007a; Laubenberger and Rowan 1982). Yet with the increase of complexity in cartographic technique, the generation, influence and control of GI became the exclusive pursuit of the professional, utilising skills and equipment outside the reach of the average hobbyist (Crone 1968; Haklay and Weber 2008).

Arguably, one of the most important developments in cartography came in 1983 when U.S. President Ronald Reagan signed a directive that allowed civilian access to the military Global Positioning System (GPS: Pellerin 2006). With a GPS tracker, an amateur volunteer could (at a low cost and with minimal operational knowledge) know the exact location of points of interest (e.g., phone boxes, pubs, traffic lights etc.) or the course of a path with the same precision as a professional cartographer.

Moving forward 22 years, the 2005 conference titled *Web 2.0* was a landmark event in the history of technology. Here the concept of dynamic interactivity was heralded as the new life of the internet over the old *web 1.0*; a network of sites that are visited, seen, but rarely changing (Tapscott and Williams 2008). Rather than proposing a new generation of technologies, O'Reilly (2005) described *Web 2.0* as a term for a loose collection of technologies and web based applications which:

1. Treats the web as a platform for services and participation,
2. Harnesses the collective intelligence of the crowd, and not just developers,
3. Relies on data richness and completeness to prove advantage over competition,
4. Are based on lightweight technologies which may be exploited by the home developer,
5. Provide continual updates and upgrades to web services,

6. Span multiple platforms (not just PC, Mac or select mobile devices)
7. Provide a rich user centred experience.

The significance of this was not the naming of the phenomenon, but recognition that *lead-users* and *developers* were moving away from a static hierarchical *Design and Use* model towards a *Use Centred Design* model. One of the movements occurring online, which prompted O'Neil to create the term Web 2.0, was that of taking geo-located data from various online locations and combing it with the newly formed *digital earths;* such as Google Maps. The result has come to be known as Neogeography (Turner 2006); commonly termed a *mashup*.

Driven by the ability to know the precise location of any point on the earth's surface with a relatively cheap GPS reader, and being able to dynamically share data in interactive ways never before possible, the mashup began evolving. GI products began taking in data not only from trained professionals, but also from untrained amateurs and the modern scene of cartography was formed. Rather than being purely for enthusiasts, these volunteer generated maps started permeating society, leading Goodchild (2007b) to coin the term Volunteered Geographic Information (VGI). Despite all these advances, Idris et al. (2011a) commented that *"there is little guidance for map mashup developers on how to design a good map that considers the quality elements before placing and publishing the data on the map"*.

This book explores the roles in which volunteered and professional information play within neogeography from a human factors perspective. The unique advantages of each information type are considered alongside how they may be utilised to create products and services delivering highly functional, efficient and satisfying experiences to their users.

1.2 The Fundamental Issues

1.2.1 Neogeography, Volunteers and Users

Web 2.0 in cartography first entered popular consciousness in 2005 with Paul Rademacher's *housemaps* website. This overlaid rental listings from the online classified-ad service Craigslist (http://www.craigslist.org) onto the recently released Google Maps (Tapscott and Williams 2008). Since its creation this process has been named neogeography (or more commonly, a *mashup*). Although first defined in its modern sense by Turner (2006), neogeography is possibly best defined by Tuchinda et al. (2008) as:

> A web application that integrates data from multiple web sources to provide a unique service, involves solving multiple problems, such as extracting data from multiple web sources, cleaning it, and combining it together.

Crucially, the advent of the neogeography opened the door to the distribution of GI created by *largely untrained volunteers* (Haklay et al. 2008). Goodchild

1.2 The Fundamental Issues

(2007a) phrased this phenomenon as *Volunteered Geographic Information* (VGI), referencing the complete or partial inclusion of volunteered information in mashups. As noted by Pultar et al. (2009), VGI can come in many different forms (e.g. restaurant reviews, travel logs, or geo-tagged photos), but in order to use any VGI for analysis and visualization in a Geographic Information System (GIS) it must be in a proper geospatial data format. While this has allowed for an in depth interaction between multiple information sources previously too complex to comprehend, Al Bakri and Fairbairn (2011) presented a series of new and previously unmet challenges to both the GI professional and the end user including accuracy, data integration, quality, region of geographic description and information attributes.

Many of the issues which may be associated with Web 2.0, Neogeography and VGI have a long standing presence in academia. For example Bédard (1986) brought attention to *meta-uncertainty* (uncertainty about uncertainty) and *uncertainty absorption* to describe the financial risks associated with providing/using spatial data. This, as noted by Devillers et al. (2010), is a fundamental concern when dealing with the new questions raised by the arrival of spatial data mashups and VGI.

Coote and Rackham (2008) commented that in the wider picture of Geographic Information (GI), two key principles are *"understanding the users' requirements"* and *"being able to assess the fitness for purpose of data and systems in an appropriate context"*. In a similar vein, Harding et al. (2009) called for a better understanding of users of VGI in terms of:

1. Which users/personas need to be understood for digital GI products to be considered usable;
2. How are existing products and formats used, by whom and for what purposes;
3. What has changed and why over the history of digital GI use, when comparing producer selected formats to user selected formats;

Considering the relation of VGI to other participation projects, Tulloch (2008) commented that for VGI to become widely accepted within the GIS field, the wider elements which contextualise the phenomenon must be understood. The comment was somewhat echoed by Goodchild (2008a) in his call for clearly defined limits of how personal VGI may be used within the wider ranges of society. Building upon these themes, Feick and Roche (2010) highlighted the question of whether the emergence of VGI alters our understanding of what constitutes GI, the way users may *value* data and how value may be understood and determined in a concept with zero transaction or delivery cost. Ultimately, at the outset of this book the geographic, cartographic, computer science and information science perspectives on the *worth* of VGI had largely been addressed as to *if* VGI can be used within neogeography. What was however unknown was how users of neogeography react to, perceive and value VGI, and if its use is beneficial or detrimental to the utility and usability of the products.

1.2.2 Users of Volunteered Information

Within the context of GI Science and spatial analysis, VGI has been shown to be *"more than accurate enough"* in its spatial positioning and content to be used alongside or instead of PGI (Haklay et al. 2009). However, the reaction of users to VGI, how they perceive it, and its effect on their activities is currently unclear. The importance of this is not the representation of the current state of VGI, but the potential level of accuracy and utility which VGI may achieve with sufficient development and contribution. Both Elwood (2008) and Zielstra and Zipf (2010) proposed that both VGI and PGI pose specific advantages and disadvantages for the *end user*, suggesting that no single information type may fulfil all of a user's requirements. It is therefore important to consider the role that the users of VGI have on its presentation, use and perception.

Questioning the importance of data quality in neogeography, Coote and Rackham (2008) commented that neogeography (and VGI) pose a distinct paradigm shift within the world of GIS:

> For those of us who have been around the industry for a while and have lived through various "paradigm shifts" observe that there are some underpinning principles that have been important throughout. Two of these principles are to (i) understand the users' requirements and (ii) be able to assess the "fitness for purpose" of data and systems in that context.

Therefore, understanding the users of VGI and neogeography is essential. Without the knowledge of (1) who the users are and (2) their *cognitive, behavioural* and *attitudinal* characteristics, then attaining user requirements for usability design is an impossible task (Gould and Lewis 1985).

Coleman et al. (2009) highlighted that although empirical research into the *contributors* and *contributions* of open source projects has been conducted, the volunteers' motivations still need to be understood alongside the relative quality of their output (Benkler 2002; Krishnamurthy 2002; Raymond 1999).

Since the advent of Web 2.0 and neogeography, GIS tools and applications on our home and work computers (e.g. laptops, tablets, smart phones, etc.) have entered the daily lives of millions around the world (Goodchild 2008b; Tapscott and Williams 2008). Predictions for future use point to widening involvement of GIS in our everyday life, with increasing levels of sophistication and complexity. One example of this is the Living Earth Simulator project, which aims to produce a *Digital Earth* (Gore 1998), collecting data from billions of sources and aiming to create a simulator that can replicate everything happening on earth (Morgan 2010). The prominence and ubiquity of such systems in today's society is best summed up by the comments of Google Earth founder John Hanke (2007) who stated that *"it is staggering to think that Google Earth and Google Maps were only introduced in the summer of 2005"*.

Although such developments carry much weight and prestige within the literature, Haklay et al. (2008) have commented that despite all the advances in user centred geography, nothing is actually new: it is just online and interactive.

However what may be considered new is the distribution of *GI tools* (e.g. remote sensing via Google Earth) previously only available to Geography Professionals (Ewert and Hollenhorst 1989).

If the pursuit of cartography and GIS products is disassociated from the professional body - as called for by Livingstone (1992)—then the user may effectively become the designer and generator of their own products in a very real and effective way (Shirky 2009). There arises the question of *why* users volunteer their time to produce products not just for their personal use, but to share with others. Trogemann and Pelt (2006) reported that *"despite all available technology, people in modern societies feel more excluded from society, more isolated with respect to their communities and more disenfranchised from the system of government and democracy"*. While this may suggest the volunteer is seeking a feel of engagement through social interaction—*social intercourse* (Kanpp 1978)—through the internet, not enough is yet known in the literature to fully understand the impact of such situations. However, this should be considered in relation to the comments of Fox (2010) that the internet has levelled the *social, economic, racial* and *cultural* divides within the USA, and to a lesser degree the relationship between its citizens and the international community.

One definition of geocollaboration is of collaborative activities in which two or more individuals work together on a single task or closely related subtasks, constructing and maintaining a shared problem concept (MacEachren and Brewer 2004). If the issue of geocollaboration surrounds the user centred understanding of neogeography, then an understanding of the catalysts for conversation between individuals and groups may prove beneficial to those wishing to utilise geocollaborative systems for the benefit of their own products (e.g. Google My Maps). Currently the understanding of why these groups come together to produce highly usable results (Haklay et al. 2009) for almost no perceivable benefit is limited.

1.2.3 Data Richness of Volunteered Information

When considering VGI it may be difficult to assess whether the data has been produced to a relevant specification of accuracy and content, so the level of data richness may be highly unknown (Daft and Lengel 1986). Coote and Rackham (2008) commented that consumers want products to work above all else, with other simple attributes such as accuracy important to them, yet they may be unable to articulate such needs. The example of *"where are the best pubs along the route"* was given by Coote and Rackham (2008) as a simple scenario that highlights how to the user the most important factor is the information directly relevant to their need, whereas other information such as phone boxes, village greens and corner shops may be interesting, yet irrelevant. The issue that arises here is *the degree to which the information is relevant to the context of* use (Coote and Rackham 2008).

Keen (2007) vocally attacked the notion of user generated content and Web 2.0 as empowering the user's creativity, yet producing overall less satisfactory outcomes of low data richness. However, Tapscott and Williams (2008) refuted this as allowing small organisations or individuals to gain an equal platform with the established professional, increasing the talent for users to choose from. Similarly, Hall (2007) reported that Google Earth's technology chief [Michael Jones] believed that individuals volunteering data creates a convergence of truth, since each contribution represents a portion of truth. In addition to this, Jones insisted that those local to the information have a stake in its accuracy. However Haklay et al. (2008) commented that the distribution of contributions over a national (UK), continental and global level—described as data richness—is currently unknown.

OpenStreetMap founder Steve Coast (report in Black 2007; Haklay and Weber 2008) commented that *"nobody wants to [contribute VGI about] council estates"*, creating a patchwork geography with important areas missing due to contributor bias. Such an uneven spread of focus from crowd sourced projects is not new. This is highlighted by Gilmartin and Lloyd (1991) that *"there is higher interest in events and geography that are local to the user, relative to faraway places"*. What is unknown here is to what impact a *patchwork spread* of VGI and data richness will have on the end users experience of using the information.

1.2.4 Trust in Volunteered Information

Ahituv et al. (1998) commented that *"the real value of information is derived from comparative measuring of differences in a decision maker's behaviour when he or she is provided with the different information sets"*. In practice, individuals typically *search* for and *use* information, they make choices whether to accept or reject discovered sources, and derive value from information based on its relevance to the task at hand (Tóth and Tomas 2011). Within this *use* situation, *trust* in the information being utilised becomes a very important aspect to the user.

Harvey (2003) described trust as being an expression of a user's underlying confidence; be it rational or irrational. Additionally, Harvey commented that trust in GIS is closely related to the users understanding of the technology with which the information is delivered. Similarly Goodchild et al. (1998) reported that the development of an understanding of trust in GI is complementary to addressing technological barriers in applications.

The subject of trust in *VGI* has yet to be directly addressed in the published literature. However, a large body of research has been generated on the issue of trust in *traditional* GI. On this, Goodchild (2008b) commented that *"if something appears to be in the wrong place would you trust it?"* Contextualisation is provided by the remarks of Kneale (2003) that *"most geographic data are noisy, imprecise, inconsistent, and may also be biased. The trick is to recognise sources of error"*. Similarly, Crampton (2010) remarked that a user must consider critically the *"truth claims of maps and GIS"* and that *"knowledge is not 'out there' but is created and then is privileged by being divided between truth and falsity"*.

Harvey (2003) commented that trust can be seen as a relationship between two parties, and is scalable in its nature. Of this Harvey counted existing social, political and professional relationships between bodies as factors which increase the level of trust in the GI being provided. An example of this was given as a government body in the U.S.A finding it easier to build a relationship of trust in GI from another U.S.A based government body (i.e. the National Spatial Data Infrastructure: NSDI) than a further removed non-governmental body. However, since trust is a personal construct in the relationship between the user and provider, it is expected that trust issues in VGI should mirror that of traditional geography.

In the literature there is no dispute that the level of trust the user has in the information they are using is important. However, what is less clear is what factors influence the user to perceive the information they are using as trustworthy enough for their given needs?

1.3 General Aim of Book

The overall aim of this book is to address the issue of how *VGI* can be combined with *PGI* to satisfy the information search requirements of consumer-users via highly usable mashups. Firstly, this required the development of an understanding of the way different users perceive VGI and PGI in terms of its benefits to their activities and information needs. Secondly, the benefits that VGI may bring to the user experience of a mashup (which cannot be attained through the use of PGI) needed to be understood. In order to achieve this, a user centred design perspective was implemented throughout the research.

References

Ahituv N, Igbaria M, Sella A (1998) The effects of time pressure and completeness of information on decision making. J Manag Inf Syst 15(2):153–172

Al Bakri M, Fairbairn D (2011) User generated content and formal data sources for integrating geospatial data. In: Proceedings of the 25th international cartographic conference and the 15th general assembly of the international cartographic association. Palais des Congres, Paris, France: ICC

Bédard Y (1986) A study of data using a communication based conceptual framework of land information systems. Can Surveyor 40:449–460

Benkler Y (2002) Coase's Penguin, or, Linux and the nature of the firm. Yale Law J 112(3):367–445

Black N (2007) OpenStreetMap—Geodata collection for the 21st century, 2010(Sept 27th). Available at: http://www.slideshare.net/nickb/nick-black-openstreetmap-geodata-collection-for-the-21st-century

Coleman DJ, Georgiadou Y, Labonte J (2009) Volunteered geographic information: the nature and motivation of producers. Int J Spatial Data Infrastructures Res 4:332–358

Coote A, Rackham L (2008) Neogeography data quality—is it an issue? In: Holcroft C (ed) Proceedings of AGI Geocommunity'08. Stratford-Upon-Avon, UK: Association for Geographic Information (AGI), p. 1. Available at: http://www.agi.org.uk/SITE/UPLOAD/DOCUMENT/Events/AGI2008/Papers/AndyCoote.pdf

Crampton JW (2010) Mapping: a critical introduction to cartography and GIS. Wiley-Blackwell, England

Crone GR (1968) Maps and their makers: an introduction to the history of cartography 4th edn. In: East WG (ed). Hutchinson, London

Daft RL, Lengel RH (1986) Organizational information requirements, media richness and structural design. Manage Sci 32(5):554–571

Devillers R et al (2010) Thirty years of research on spatial data quality: achievements, failures, and opportunities. Trans GIS 14(4):387–400

Elwood S (2008) Volunteered geographic information: future research directions motivated by critical, participatory, and feminist GIS. GeoJournal 72:173–183

Ewert AW, Hollenhorst S (1989) Testing the adventure model: empirical support for a model of risk recreation participation. J Leis Res 21(2):124–139

Feick R, Roche S (2010) Valuing volunteered geographic information (VGI): opportunities and challenges arising from a new mode of GI use and production. In: Poplin A, Craglia M, Roche S (eds) GeoValue 2010 proceedings: 2nd workshop on value of Geoinformation. Hamberg, DE, Geovalue, HafenCity University Hamberg, p 67

Gilmartin P, Lloyd R (1991) The effects of map projections and map distance on emotional involvement with places. Cartogr J 28(2):145–151

Goodchild MF (2007a) Citizens as sensors: the world of volunteered geography. GeoJournal 69(4):211–221. Available at: http://www.springerlink.com/content/h013jk125081j628/

Goodchild MF (2007b) Citizens as voluntary sensors: spatial data infrastructure in the world of web 2.0. Int J Spatial Data Infrastruct Res 2:24–32

Goodchild MF (2008a) Commentary: whither VGI? GeoJournal 72(3):239–244

Goodchild MF (2008b) Spatial accuracy 2.0. In: Zhang J-X, Goodchild MF (eds) Proceeding of the 8th international symposium on spatial accuracy assessment in natural resources and environmental sciences. World Academic Union, Shanghai, pp 1–7

Goodchild MF, Egenhofer MJ, Fegas R (1998) Interoperating GISs: report of the specialist meeting, Santa Barbara, CA, NCGIA Varenius Project

Gore A (1998) The digital earth: understanding our planet in the 21st century. Aus Surveyor 43(2):88–91. Available at: http://www.tandfonline.com/doi/abs/10.1080/00050326.1998.10441850?journalCode=tjss18

Gould JD, Lewis C (1985) Designing for usability: key principles and what designers think. Commun ACM 28(3):300–311

Haklay M, Weber P (2008) OpenStreetMap: user-generated street maps. IEEE CS, October-De, pp 12–18

Haklay M, Singleton A, Parker C (2008) Web mapping 2.0: the Neogeography of the Geoweb. Geogr Compass 2(6):2011–2039

Haklay M, Ather A, Zulfiqar N (2009) Beyond good enough? spatial data quality and OpenStreetMap data. Slideshare, 2009(July 29th). Available at: http://www.slideshare.net/mukih/beyond-good-enough-spatial-data-quality-and-openstreetmap-data (Accessed April 3, 2013)

Hall M (2007) On the mark: will democracy vote the experts off the GIS Island? 2009(Nov 18th). Available at: http://www.computerworld.com/s/article/299936/Will_Democracy_Vote_the_Experts_Off_the_GIS_Island_

Hanke (2007) Distinguished Innovator Lecture, University of California, Berkeley, 19 November 2009

Harding, Jenny, Sharples et al. (2009) Usable geographic information—what does it mean to users? In: Proceedings of the AGI GeoCommunity '09 Conference, Stratford-Upon-Avon, UK: AGI GeoCommunity

References

Harvey F (2003) Developing geographic information infrastructures for local government: the role of trust. Can Geogr 47(1):28–36

Idris NH, Jackson MJ, Abrahart RJ (2011) Colour coded traffic light labeling: a visual quality indicator to communicate credibility in map mash-up applications. In: Presented at the international conference on humanities, social sciences, science & technology (ICHSST), Manchester, UK, pp 1–7. Available at: http://www.geoinfo.utm.my/geoinformatic/geoinformatic publications/2011/Nurul_Idris_ICHSST_new.pdf

Kanpp ML (1978) Social intercourse: from greeting to goodbye. Allyn and Bacon, Boston

Keen A (2007) The cult of the amateur. Nicholas Brealey, Finland

Kneale PE (2003) Study skills for geography students: a practical guide. Hodder Arnold, England

Krishnamurthy S (2002) Cave or community? an empirical examination of 100 mature open source projects. First Monday 7(6):1–12. Available at: http://papers.ssrn.com/sol3/papers.cfm?abstract_id=667402

Laubenberger F, Rowan S (1982) The naming of America. Sixt Century J 13:91–113

Livingstone DN (1992) In defence of situated messiness: geographical knowledge and the history of science. GeoJournal 26(2):228–229

MacEachren AM, Brewer I (2004) Developing a conceptual framework for visually-enabled geocollaboration. Int J Geogr Inf Sci 18(1):1–34

Morgan G (2010) Earth project aims to "simulate everything. 2010(Dec 28th). Available at: http://www.bbc.co.uk/news/technology-12012082

O'Reilly T (2005) What is web 2.0: design patterns and business models for the next generation of software. oreilly.com, 2010(Dec 27th), pp 1–5. Available at: http://oreilly.com/web2/archive/what-is-web-20.html [Accessed April 2, 2013]

Pellerin C (2006) United States updates global positioning system technology. america.gov, 2011(Jan 2nd). Available at: http://www.america.gov/st/washfile-english/2006/February/20060203125928lcnirellep0.5061609.html [Accessed April 2, 2013]

Pultar E et al (2009) Dynamic GIS case studies: wildfire evacuation and volunteered geographic information. Trans GIS 13:85–104

Raymond E (1999) The Cathedral and the Bazaar. Knowl, Technol & Policy 12(3):23–49

Shirky C (2009) How cellphones, Twitter, Facebook can make history. In: Frawley Bagley E (ed) TED@State. TED Talks, Washington. Available at: http://www.ted.com/talks/clay_shirky_how_cellphones_twitter_facebook_can_make_history.html

Tapscott D, Williams AD (2008) Wikinomics: how mass collaboration changes everything. Atlantic Books, UK

Tóth K, Tomas R (2011) Quality of geographic information—simple concept made complex by the context. In: Proceedings of the 25th international cartographic conference and the 15th general assembly of the international cartographic association. Palais des Congres, Paris, France: ICC

Trogemann G, Pelt M (2006) CITIZEN MEDIA—technological and social challenges of user driven media. In: Proceedings of the BroadBand Europe conference. Geneva, Switzerland: BroadBand Europe, pp 1–6. Available at: http://interface.khm.de/wp-content/uploads/2008/10/06-bbeurope.pdf

Tuchinda R, Szekely P, Knoblock CA (2008) Building mashups by example. In: Proceedings of the 13th international conference on Intelligent user interfaces. Canary Islands, ACM, pp 139–148. Available at: http://dl.acm.org/citation.cfm?id=1378792

Tulloch DL (2008) Is VGI participation? From vernal pools to video games. GeoJournal 72:161–171

Turner AJ (2006) Introduction to Neogeography, eBook: O'Reilly Media. Available at: http://shop.oreilly.com/product/9780596529956.do?CMP=OTC-KW7501011010&ATT=neogeography#

Zielstra D, Zipf A (2010) A comparative study of proprietary Geodata and volunteered geographic information for Germany. In: Painho M, Santos MY, Pundt H (eds) Geospatial thinking: proceedings of the 13th AGILE international conference on geographic information science. Guimarães, Portugal, AGILE, pp 1–15

Chapter 2
A Framework of Neogeography

2.1 Introduction

Within the current literature, confusion exists as to the terminology used for the various technologies, innovations and phenomenon associated with VGI. This is best highlighted by Elwood (2008) in that *these developments [in geotagging data] have been referred to with a plethora of terms, including neogeography... web mapping... volunteered geographic information... ubiquitous cartography... and wiki-mapping*. This extensive list is added to by Crampton (2008) with *Spatial Media*, *Locative Media*, *Spatial Crowdsourcing*, *Geocollaboration* and *Map Hacking*. Suggesting an explanation for this, Tulloch (2008) suggests that initial islands of research producing unique or proprietary vocabulary may introduce *buzzwords* which suit their cause, yet die out over time. As Crampton (2008) commented, the [neogeographic] situation has from its birth been both increasingly important and *interestingly messy*, with its confusing terminology being linked with the emergence of the Web 2.0 and Neogeographic phenomenon itself (Das and Kraak 2011).

The confusion highlighted by Elwood (2008) and Crampton (2008) is further underlined in how neither goes on to distinguish between these various definitions. Neither do they present a distinction between the types of data type or technique being described. The lack of agreement on terms by these and other authors (Coote and Rackham 2008; Haklay et al. 2008; Shin 2009) highlights the lack of consensus in terminology, leading to multiple authors using various different phrases to describe the same thing. In order to avoid such detrimental mistakes within this book, the following must be achieved:

- Set out the *true* definitions of the terms related to neogeography, providing a consensus for this book and hopefully further work.
- Discuss the way in which the different elements of neogeography interact with one another, providing a framework on which the information types in this book shall be based.
- Develop a framework of neogeography so neogeographic projects may be effectively compared and contrasted through this book.

2.2 Background Literature

2.2.1 The Nature of Neogeography

Often in the literature, the terms *Neogeography, Mashup* and *VGI* are substituted for other terms such as Public Participatory GIS (Aberley and Sieber 2002) or Geoweb (Haklay et al. 2008). This is often without full justification for the change, and without full and proper definitions. Although adding to the general confusion of what is VGI, this helps to suggest that the different names given to VGI and Neogeography need to be addressed and fully defined, allowing their appropriate use through common understanding.

One example includes the comments by Idris et al. (2011, p. 120) who claimed *"neogeography relies on user generated content that is locationally tagged"*. Although Idris et al. were correct on the reliance of locational data within neogeography, their statement that user generated content (VGI) is a necessary component to Neogeography was incorrect.

While the term *neogeography* has been used in various forms from at least 1944 (Miller and Miller 1944), it was Turner (2006, p. 2) who cemented the term in the form it is used and understood within this book:

> Neogeography means "new geography" and consists of a set of techniques and tools that fall outside the realm of traditional GIS, Geographic Information Systems. Where historically a professional cartographer might use ArcGIS, talk of Mercator versus Mollweide projections, and resolve land area disputes, a neogeographer uses a mapping API like Google Maps, talks about GPX versus KML, and geotags his photos to make a map of his summer vacation.

According to his description, neogeographic systems may exist and function in the fullest sense while relying only on professional information sources; see Fig. 2.1. However, the need to present the disconnection between neogeography, VGI and PGI denote a degree of further explanation is required in order to fully define the terminology relevant to this book.

To understand neogeography this chapter deals with the various elements of the phenomenon, with each taxonomy list relating to one particular element of the phenomenon. For simplicity, these elements are referred to as:

- *Data Generation Aspect*—People, either volunteers or professionals creating raw data; VGI or PGI.
- *Neogeographic Aspect*—Combining geo-data with a form of map to produce a mashup.
- *User Aspect*—Referring to any group or individual who takes the product of the neogeographic element and utilises it in some way.

The interaction between these three elements is highlighted in Fig. 2.1 above.

Figure 2.1 highlights how neogeography is the process of combining geo-data with maps to create mashups, whereas VGI and PGI are simply the creation of one form of data. It is important to mark the distinctions between VGI and PGI. VGI is

2.2 Background Literature

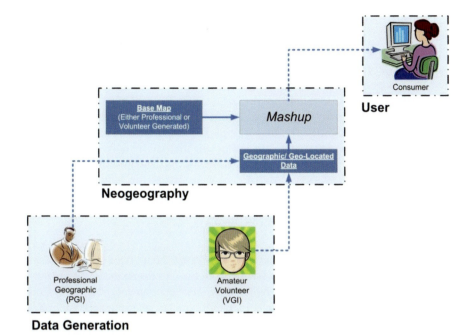

Fig. 2.1 Elements constructing the neogeographic phenomenon

essentially geographic information created by largely untrained amateur volunteers (Haklay and Weber 2008). In defining VGI, Goodchild (2007a) opened up the scope of *geographic objects* that could be described through volunteered means to be *"not confined to traditional geographic identifiers such as trees and streets but to any data where a geospatial element is present"*. However, it does not exclude professionals or organisations from contributing. This has resulted in projects where the quality in terms of positional accuracy and data richness of VGI projects may outreach that of similar PGI projects (Haklay 2010b). However, while a professionally trained person may contribute to a VGI project, it would be predominantly as a hobby using the same tools as the amateur volunteer, and without any privileges or advantage.

Further to the naming of the information based on the professionalism of the author is the issue of how the geographic objects are being described in a more general sense. In the context of consumer products, Zeithaml (1988) regarded the elements of *price, quality* and *value* as important descriptors for the ways different people interact with information. However, according to Zeithaml (1988) and Sheridan (1995), the perspectives of *quality* and *value* are relative to both application and use. This suggests that utilising user perceptions of information may not necessarily be the best way to categorise projects within the framework. This is because a user may perceive two very different mashups (containing different data and use characteristics) as being equal in utility, efficiency and satisfaction.

Additionally, price is not necessarily a good descriptor either, due to the non-traditional business model usually applied to neogeographic products (Tapscott and Williams 2008).

To take a more user centred design perspective, mashups and neogeography are tools utilised by users to achieve their goals and to create products specific to their personal requirements. Das and Kraak (2011). gave the example that a user can create a map *showing* all local fitness centres; presenting collected data. Alternatively, a user may use the same map to *explore* local fitness centres. This creates two distinct design opportunities since although the data required by both user groups is the same, their use and relationship with the data are different.

2.2.2 Issues with Current Taxonomies

From a GIS perspective, Grimshaw (1996, 1992) highlighted how previous taxonomies had oversimplified the viewpoints of the GIS discipline and assumed a static technological infrastructure, rather than one that changes over time. Consequently, Grimshaw (1996) produced a more complex and overarching framework consisting of Management Strategy, Technology and Decision. Bai et al. (2009) noted that this framework is rooted in the key concepts of information systems, yet departs from the concrete functionalities, specific communication protocol definitions and expected usage scenarios within geospatial sciences. This has in turn prevented it from being properly utilised. However, the largely dynamic, unstructured and *anarchic* nature of neogeography (Budhathoki et al. 2008) suggests that the production of a framework along a similar approach may prove more useful than when applied to the more rigid platforms in GIS. Additionally, while a justification for using the framework of Grimshaw (1996) may be possible, the dimensions do not sit comfortably within the neogeographic literature. Therefore a more appropriate and accessible framework is required to fulfil the need for a relevant classification system for neogeographic projects.

Coleman et al. (2009) produced a series of models relating specifically to VGI, characterising, amongst other things, the spectrum of contributors, characteristics of use, motivations to contribute and the institutional requirements. Whilst interesting and insightful, their disjointed nature (i.e. the lack of connection and integration between the models) makes them difficult to use in an overarching framework. A more recent attempt at classifying VGI within a taxonomy was provided by Cooper et al. (2011), who identified dimensions of VGI and Neogeography as being:

- The continuum of responsibility for determining the specification of the data.
- The classification of data from base (e.g. streets networks) to Points of Interest (POIs).

A weakness of the framework is that the presentation of the framework is largely inaccessible due to its reliance on unconventional terminology (e.g.

custodian and POI not commonly used in neogeographic literature) and its basis on informality. This is a theoretical perspective at odds with the lack of universal standards of procedures across the spectrum of neogeography, constantly changing to fit the desire or needs of the producers. Additionally Cooper et al. (2011) combined both neogeographic project with GIS phenomenon (e.g. *tracks4africa* and *PPGIS*), which while interesting from a taxonomy perspective are two incompatible concepts within a single framework.

2.3 A Framework of Neogeography

2.3.1 A Terminology of Neogeography

The provision of a terminology is necessary in order to overcome the potential confusion amongst neogeographic creators and those wishing to discuss neogeographic phenomenon. Although a detailed overview of definitions relative to this book is provided in the glossary at the start of this volume, it is necessary to highlight the key terms this taxonomy related to; see Table 2.1.

In the advent of neogeography, Al Bakri and Fairbairn (2011) presented a series of new and previously unmet challenges to the world of geo-information, focusing on accuracy, data integration, quality, region of geographic description, and information attributes. This list may be added to by considering more traditional metrics of GI; quality (Devillers et al. 2010), accountability (Coleman 2009) and data standards (Brando et al. 2011).

Although research has demonstrated VGI to be able to produce information to the same *quality* as PGI (Haklay et al. 2009; Haklay2010a, b), the optimal word here is 'able'. That being, simply because one project (e.g. OpenStreetMap) is able to produce maps as good as OS Meridian, does not mean that all are (e.g. ThePeoplesMap). While looking further into the reason for this high accuracy coming from amateur volunteers, Haklay et al. (2010) demonstrated that at least five edits from proficient persons is required to converge on a truth of high enough quality. Therefore, we may consider the degree of standardisation in how data are produced as a mechanism for achieving high quality products. While PGI sources have a long and established history of standardisation of practices (Crone 1968), VGI may be considered anarchic (Budhathoki et al. 2008). As Brando (2011) demonstrated, the way in which VGI is produced, categorised and retrieved may be standardised within a project to an efficient and effective level, there is no guarantee of such implementation. Further too this, the very concept of standardisation of VGI is alien to the anarchic mechanism of producers *doing as they will* to produce the products they desire in the way they see fit. A concern of professionals which is prevalent within the scoping study of this book is the concern for accountability and trust as derived from VGI. Due to the high degree of quality control within PGI (Goodchild 2000), this information form has been the

Table 2.1 Key definitions within the framework

Term	Definition
Geographic Information Systems (GIS)	Medyckyj-Scott and Hearnshaw (1993) described GIS as "*tools that capture, store, manage, manipulate, analyse, model and display information with respect to geographical space*"
Base Map	A raster map used within a mashup on which information is layered (Das and Kraak 2011)
Neogeography	Turner (2006) defined neogeography as "*people using and creating their own maps, on their own terms and by combining elements of an existing toolset*". In a broader research application context, Das and Kraak (2011) described this as "*the domain where users make use of geographic information*" (GI) using Web 2.0 applications
Professional Geographic Information (PGI)	While not a phrase in common use throughout the current literature, the term *Professional Geographic Information* (PGI) has been utilised within this book to make reference to geographic information not originating from volunteers; in contrast to VGI. This may be defined as structured geographic information produced by trained personnel (Fonseca and Sheth 2002), or those of able to provide detailed geographic information that can be verified and integrated at the national level (Goodchild 2007b)
Volunteered Geographic Information (VGI)	Goodchild (2007a) referred to this phenomenon as "*geographic information created by largely untrained volunteers,* which is *potentially unstructured*" (Fonseca and Sheth 2002)

bedrock of personal through to governmental actions since the creation of GI, notably in police, fire, rescue and military situations (etc.). Due to the lack of standardisation with VGI (Cooper et al. 2011; Zook et al. 2010) such equal implementation has been hampered and continues to be the alternative to PGI only when PGI is not fully available. However, quality control metrics have been, and are included within crowd-sourced projects. Examples for this range from the peer review and peer pressure of Wikipedia, through to automatic quality control filters of Tracks4Africa where contributed data must reach a minimum degree of logical consistency before it is accepted into the main data set (Cooper et al. 2011).

2.3.2 A Framework for Neogeography

The most fundamental aspect of a framework is the dimensions by which the subject matter is categorised. Within the general sense of geographic information, Coote and Rackham (2008) highlighted the four dimensions of *completeness, consistency, quality control* and *quality assurance* as key areas of concern within neogeography. While each of those points is valid, the one that stands out as most revenant to this section is quality control. This is for a variety of reasons; most notably (as highlighted above in the terminology of neogeography) that the amount

2.3 A Framework of Neogeography

of quality control put in place is of high concern to a variety of users. Additionally, Goodchild (2008) highlighted this as one of the greatest challenges facing VGI, and Zeithaml (1988) and Sheridan (1995), placing quality as relative to both application and use. Furthermore, the conditions of completeness, consistency and quality assurance can either be considered as temporary states (i.e. the data set may become more complete over time), or can be addressed through proper quality control.

Alexander and Tate (2005) cited *authority, accuracy, objectivity, currency* and *coverage* as the key factors in assessing the *appropriateness* of an information source to a user's information search requirements. Out of these, objectivity was selected as the most appropriate second dimension of the framework. Within a general research context, both Boudreau et al. (2001) and Janesick (2000) considered o*bjectivit*y to be one of the most crucial to the ratification of information. The remaining dimensions of appropriateness were not selected since it was not felt that their position was well enough supported in relation to VGI and its current understanding in the literature.

Because the evaluative judgement made by the user on information is comprised of *opinions, attitudes* and *beliefs* (Albaum 1997; Mizumoto and Takeuchi 2009), a need exists to quantify projects in an objective form. According to Preece et al. (2011), usually the most appropriate method of investigating the participant's response to information presented in a study is through subjective rating using Rating Scales. Tables 2.2 and 2.3 propose two Rating scales for quantifying both *quality control* measures and the level of *objectivity*.

Building on the Rating Scales of Tables 2.2 and 2.3 and evolving the approach of Cooper et al. (2011), Fig. 2.2 presents a framework for how to consider and categorise neogeographic products.

Table 2.2 Rating scale for assessing quality control

Level of quality control	Definition
1 *None*	All data entries are accepted into the data base without any control over any attributes, data cannot be edited or removed by anyone but the author
2 *Very low*	All data entries are accepted into the data base without any control over any attributes, data can be edited or removed by anyone
3 *Low*	Data may be accepted into the data base providing the minimum meta data requirements are met, data can be edited or removed by anyone
4 *Intermediate*	Data may be accepted into the data base providing the minimum meta data requirements are met, checked before being added to the system, data can be edited or removed by anyone
5 *Absolute*	All aspects of data entering the system must strictly comply to a pre-specified standards, data checked before being added to the system, edited and/or removed by any other person in the system with authority

Table 2.3 Rating scale for assessing objectivity

Level of objectivity	Definition
1 Totally subjective	No way of verifying any of the data through quantitative measurements, can only come from users forming their own opinions
2 Mostly subjective	Most data has to come from users forming their own opinions, although a degree of quantitative measurement is required
3 Equally subjective and objective	All data can be achieved through either qualitative measurement, or through users forming their own opinions
4 Somewhat objective	Most data has to come from quantitative measuring methods, although some data should come from users forming their own opinions
5 Totally objective	All of the data can only be achieved through quantitative measurements

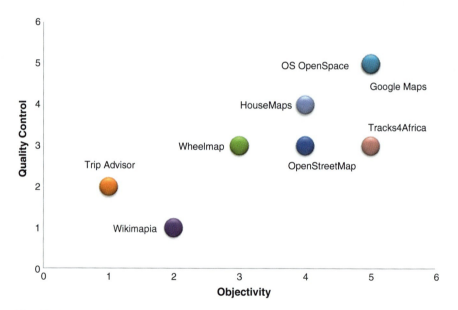

Fig. 2.2 A proposed framework for neogeographic products

2.4 Discussion

The purpose of presenting this framework through a scatter graph is to allow a simple way to visualise how similar or dissimilar various projects may be, as judged by the objectivity and quality elements. For example, within Fig. 2.2 the close proximity of *Ordnance Survey, OpenStreetMap* and *Google Maps* suggests that while their focus may be different, they may be considered alongside each other and be categorised together; even though they are VGI and PGI projects.

2.4 Discussion

However, OpenStreetMap is a very distant from Wikimapia, since OpenStreetMap is a project producing an objective map of features (e.g. roads, buildings, post boxes, etc.) while Wikimapia produces a subjective layer of descriptions on top of an existing map (e.g. 'the pub in *this* part of the map is The Red Lion…. here is why I think it is very good'). This means these two ventures should be categorised as different forms of neogeographic products; despite both products being VGI based.

An interesting outcome from Fig. 2.2 is how when the various projects are considered against the categorisation of Tables 2.2 and 2.3, there appears to be a correlation between objectivity and quality control. Although the causal link between objectivity and quality control has been disputed (Stiles 1993), it does provide an interesting insight. If a neogeographic project seeks to capture rich user experiences about locations (e.g. the best spot on an island to watch the sun go down) then the framework suggests that low-level quality control is suitable for capturing such objective information. Similarly, to produce a mashup that describes geo-located information in a highly reliable fashion, information about locations (e.g. positions of post boxes), then a high degree of quality control is appropriate.

As highlighted previously, prior to this framework there did not exist a simple, effective and easily understood framework by which to consider different neogeographic and GIS products. Potential uses of such a framework could be considered as follows:

Selection of a product for use—This framework could be used to assess the degree to which new neogeographic products should be thought of in terms of their accountability and ability to provide meaningful, descriptive information to the user. This is particularly relevant when the information is to be used in highly sensitive situations where a degree of risk is involved; e.g. information for hospital paramedics.

Understanding neogeography in research—From a research perspective the framework outlines how although a large collection of projects can be considered neogeography, they can be very different. Therefore, future research should not look to treat (for example) Wikimapia and Tracks4Africa as the same since they fall into different categories of neogeography. However, comparing them as two different types of products, and understanding that their nature is very different may lead to a deeper and more useful investigation into how neogeography is used in society. Utilising the framework in this way would help to reduce the confusion in how neogeography is discussed in the literature.

A framework for quality control—As proposed by Bishr and Mantelas (2008), VGI data sets could be filtered to remove instances of VGI which do not meet a pre-specified quality control metric. The categories of the framework could be employed as such a metric to automatically assess the suitability of individual VGI contributions. For example, to produce a VGI contribution framework that would allow the end product to occupy the same space within the framework as OS OpenSpace, data would have to comply to a strict metadata structure, and be verified by others before it is published. This would allow mashups of (to a degree) certified accountability to be developed from sources which in their complete state offer a wide variation in quality which make them unsuitable.

Development of new products—One of the most important aspects of any innovative new product or service is its unique attributes and ability to satisfy a currently unmet user need. By considering current neogeographic products alongside this framework the niches yet to be exploited may understood, making this framework a useful tool for designers.

2.5 Conclusion

This chapter has helped address the research question of *what is VGI and how does it differ from PGI* by producing a detailed terminology and a working framework based on two of the key variables in the field of neogeography; *quality control* and *objectivity* of the information. Additionally utilising the framework allows for a useful way to discuss the differences and similarities between projects. As well as addressing the research aims, research within this book will aim to produce sufficient evidence to critically consider the dimensions that constitute this framework for their appropriateness and relevance to the user.

References

Aberley D, Sieber R (2002) About PPGIS In: Developed at first international PPGIS Conference held by URISA at Rutgers University, URISA, New Brunswick, New Jersey. Available at: http://www.ppgis.net/ppgis.htm. Accessed 3 April 2013

Al Bakri M, Fairbairn D (2011) User generated content and formal data sources for integrating geospatial data. In: Proceedings of the 25th international cartographic conference and the 15th general assembly of the International Cartographic Association, ICC, Palais des Congres, Paris

Albaum G (1997) The Likert scale revisited: an alternate version. J Market Res Soc 32(2):331–348

Alexander J, Tate M (2005) Evaluating web resources. Wideneer University, 2011 (March 25th). Available at: http://www.widener.edu/about/campus_resources/wolfgram_library/evaluate/original.aspx. Accessed 3 April 2013

Bai Y, Di L, Wei Y (2009) A taxonomy of geospatial services for global service discovery and interoperability. Comput Geosci 35(4):783–790

Bishr M, Mantelas L (2008) A trust and reputation model for filtering and classifying knowledge about urban growth. GeoJournal 72(3–4):229–237

Boudreau M, Gefen D, Straub D (2001) Validation in IS research: a state-of-the-art assessment. MIS Q 25(1):1–16

Brando C, Bucher B, Abadie N (2011) Specifications for user generated spatial content. Advancing Geoinf Sci Changing World 1(6):479–495

Budhathoki NR, Bruce B (Chip), Nedovic-Budic Z (2008) Reconceptualizing the role of the user of spatial data infrastructure. GeoJournal 72(3):149–160. Available at: http://link.springer.com/article/10.1007%2Fs10708-008-9189-x?LI=true#page-2

Coleman G (2009) Code is speech: legal tinkering, expertise, and protest among free and open source software developers. Cult Anthropol 24(3):420–454

Coleman DJ, Georgiadou Y, Labonte J (2009) Volunteered geographic information: the nature and motivation of produsers. Int J Spat Data Infrastruct Res 4:332–358

Cooper AK et al (2011) Challenges for quality in volunteered geographical information. In AfricaGEO 2011. AfricaGEO, Cape Town, p 13

Coote A, Rackham L (2008) Neogeography data quality—is it an issue? In: Holcroft C (ed) Proceedings of AGI Geocommunity'08. Association for Geographic Information (AGI), Stratford-Upon-Avon, UK, p. 1. Available at: http://www.agi.org.uk/SITE/UPLOAD/DOCUMENT/Events/AGI2008/Papers/AndyCoote.pdf

Crampton JW (2008) Cartography: maps 2.0. Prog Hum Geogr 33(1):91–100

Crone GR (1968) Maps and their makers: an introduction to the history of cartography 4th ed. East WG (ed), London, Hutchinson

Das T, Kraak MJ (2011) Does neogeography need designed maps? In: Proceedings of the 25th international cartographic conference and the 15th general assembly of the international cartographic association, International Cartographic Association (ICA), Paris, pp 1–6. Available at: http://www.itc.nl/Pub/GIP/GIP-Academic-Output/GIP-Output.html?l=6&y=2011&d=GIP

Devillers R et al (2010) Thirty years of research on spatial data quality: achievements, failures, and opportunities. Trans GIS 14(4):387–400

Elwood S (2008) Geographic information science: new geovisualisation technologies emerging questions and linkages with GIScience research. Prog Hum Geogr 33(2):256–263

Fonseca F, Sheth A (2002) The geospatial semantic web, UCGIS, Leesburg. Available at: http://www.ucgis.org/priorities/research/2002researchPDF/shortterm/e_geosemantic_web.pdf. Accessed 30 Oct 2012

Goodchild MF (2000) Communicating geographic information in a digital age. Ann Assoc Am Geogr 90(2):344–355

Goodchild MF (2007a) Citizens as sensors: the world of volunteered geography. GeoJournal 69(4):211–221. Available at: http://www.springerlink.com/content/h013jk125081j628/

Goodchild MF (2007b) Citizens as voluntary sensors: spatial data infrastructure in the world of web 2.0. Int J Spat Data Infrastruct Res 2:24–32

Goodchild MF (2008) Commentary: whither VGI? GeoJournal 72(3):239–244

Grimshaw DJ (1992) Towards a taxonomy of information systems: or does anyone need a Taxi? J Inf Technol 7:30–36

Grimshaw DJ (1996) Towards a taxonomy of geographical information systems. In: Proceedings of the 29th annual hawaii international conference on system sciences. IEEE Computer Society, Maui, p 547

Haklay M (2010a) How good is volunteered geographical information? a comparative study of openstreetmap and ordnance survey datasets. Environ Plann B 37(4):682–703

Haklay M (2010b) Openstreetmap completeness for England—March 2010. Slideshare, 2011 (Jan 2nd). Available at: http://www.slideshare.net/mukih/openstreetmap-completeness-for-england-0310. Accessed 3 April 2013

Haklay M, Weber P (2008) Openstreetmap: user-generated street maps. IEEE Pervas Comput 7(4):12–18

Haklay M, Singleton A, Parker C (2008) Web mapping 2.0: the neogeography of the geoweb. Geogr Compass 2(6):2011–2039

Haklay M, Ather A, Zulfiqar N (2009) Beyond good enough? spatial data quality and openstreetmap data. Slideshare (July 29th). Available at: http://www.slideshare.net/mukih/beyond-good-enough-spatial-data-quality-and-openstreetmap-data. Accessed 3 April 2013

Haklay M, Ather A, Basiouka S (2010) How many volunteers does it take to map an area well? In Haklay M, Morley J, Rahemtulla H (eds). In: Proceedings of the GIS research UK 18th Annual Conference, University College London, pp 193–196

Idris NH, Jackson MJ, Abrahart RJ (2011) Map mash-ups: what looks good must be good? In Emma Jones C et al. (eds). In: Proceedings of the 19th GIS research UK annual Conference, GIS Research, Portsmouth, UK, p 119

Janesick VJ (2000) The choreography of qualitative research design: minutes, improvisations, and crystallisation. In: Denzin NK, Lincoln YS (eds) Handbook of qualitative research. Sage Publications, Thousand Oaks, pp 379–399

Medyckyj-Scott D, Hearnshaw HM, (1993) preface. In Medyckyj-Scott D, Hearnshaw HM (eds) Human factors in geographical information systems, Belhaven Press, London, pp xvii–xx

Miller ME, Miller IL (1944) Encyclodedia of bible life, 4th edn. Harper and Brothers Publishing, New York

Mizumoto A, Takeuchi O (2009) Comparing frequency and trueness scale descriptors in a Likert Scale questionnaire on language learning strategies. JLTA J 12:116–136

Preece J, Rogers Y, Sharpe H (2011) Interaction design: beyond human-computer interaction. Wiley, New York

Sheridan TB (1995) Reflections on information and information value. IEEE Trans Syst Man Cybern 25(1):194–196

Shin ME (2009) Democratizing electoral geography: visualizing votes and political neogeography. Polit Geogr 28:149–152

Stiles WB (1993) Quality control in qualitative research. Clin Psychol Rev 13(6):593–618

Tapscott D, Williams AD (2008) Wikinomics: how mass collaboration changes everything. Atlantic Books, UK

Tulloch DL (2008) Is VGI participation? From vernal pools to video games. GeoJournal 72:161–171

Turner AJ (2006) Introduction to neogeography, eBook: O'Reilly Media. Available at: http://shop.oreilly.com/product/9780596529956.do?CMP=OTC-KW7501011010&ATT=neogeography

Zeithaml VA (1988) Consumer perceptions of price, quality, and value: a means-end model and synthesis of evidence. J Mark 52(3):2–22

Zook MA et al (2010) Volunteered geographic information and crowdsourcing disaster relief: a case study of the haitian earthquake. World Med Health Policy 2(2):7–33

Chapter 3
Scoping Study: User Perceptions of VGI in Neogeography

3.1 Introduction

Current research into Volunteered Geographic Information—VGI (Goodchild 2007a)—in the context of neogeography has revolved around the computer science perspectives of its utilisation for technical benefit (University of Heidelberg 2010). Although VGI has been shown to be *more than accurate enough* (Haklay 2010a) in its spatial positioning, the reaction of users to VGI, how they perceive it, and its effect on their lives is less clear.

While various authors have presented a series of conceptual frameworks to the classification of users associated with neogeography and VGI (Coote and Rackham 2008; Budhathoki et al. 2008; Sommerville 2007), the relationship between the user and their perceptions of VGI useful in a User Centred Design (UCD) context has (to date) not been covered in the published literature. In relation to the distinct lack of human factors research into VGI (Harding et al. 2009), any designer wishing to produce mashups utilising a UCD approach—and including VGI as a key data source—would be doing so without informed guidelines on how the users perceive the information they are interacting with. More importantly, it is unclear what the differences and similarities are between the perspectives of different user groups (i.e. those who are using a VGI for some purpose), and how might this effect the design of VGI inclusive mashups in the future. Consequently, a need exists to investigate the scope of users associated with VGI in order to set the theoretical foundations for a UCD understanding within this field.

3.2 Aims

The aim of this study was to better understand the phenomenon of VGI within the context of its use in neogeography. In order to tackle this, three objectives were produced:

C. J. Parker, *The Fundamentals of Human Factors Design for Volunteered Geographic Information*, SpringerBriefs in Geography, DOI: 10.1007/978-3-319-03503-1_3, © The Author(s) 2014

1. What is the nature of VGI in general?
2. What are the different characteristics of the key users?
3. How do different users perceive VGI in terms its *value* to them?

Due to the lack of published work giving a human factors perspective on neogeography and VGI, this study aimed to lay the foundations of investigation. This was then to allow the development of hypothesis and then theory in later investigations. Consequently, this study did not set out to produce a simple snapshot of user perceptions, but instead gain a detailed and useful analysis of the relevant users and their associated stakeholders.

3.3 Study Rationale

The overall rationale of this study was to understand the differences in user perception of VGI through investigating the users of different neogeographic platforms through a value framework. A series of popular map platforms were selected to produce a useful cross section of opinions relating to the overarching topic of neogeography. For each platform, appropriate users were sought and interviewed, alongside participatory observation in their activities. Through this, the study objectives laid out above were investigated. This section describes these processes alongside their rationale for the purpose of justifying the research within this chapter.

3.3.1 Selection of VGI Platforms

Due to the exploratory nature of this study, it was important that the participants reflected the diversity of opinions and practices within the wider field of VGI. Consequently, three map products were selected, describing a useful cross-section of users, technologies and attitudes.

The first map product needed to reflect the most commonly used and respected form of VGI available. OpenStreetMap (OSM) was chosen as a popular *VGI* application, where potentially untrained volunteers create and *provide free geographic data such as street maps to anyone who wants them* (OpenStreetMap 2009). Here, the main objective is the creation of the map and its associated metadata via volunteered means. OSM represents the best researched of all neogeographic products and is often used to define VGI.

The second map product needed to reflect the personal (and possibly anarchic) nature of neogeography. In line with current research into VGI creation through GIS tools (Foth et al. 2009; Miller 2006; Rinner et al. 2011) Google Maps (My Maps) was chosen as a popular *neogeographic tool* where users *create personalised, annotated, customised maps* (Google 2010). Unlike OpenStreetMap, Google

My-Maps users add *pin-points* or *poly-lines* which are then annotated with specific information.

The third map product needed to provide a perspective from the traditional/ professional side of neogeography. An additional category of participant is the traditional GIS professional. It is important to study the neogeography phenomenon relative to traditional mapping, since recent developments have not added new functionality to geographic information, but rather new approaches to geographic information distribution, usability and application development (Haklay et al. 2008). For this, Ordnance Survey was selected because of its position as the official mapping agency of the UK.

3.3.2 Investigation Overview

This chapter comprised a multi-methods investigation into the way different user communities perceive VGI in terms of its value and meaning to them. Two independent investigations were conducted, comprising participatory observation to understand the social factors and interactions between users, and semi-structured interviews for in depth investigation into user perceptions. Participants were asked to consider past and current experiences, positive and negative aspects of VGI and PGI, as well as interactions between different information types and the user community. Data was analysed through thematic analysis, with multidimensional value used a theoretic framework. Results were analysed separately, but brought together in the discussion and conclusion.

3.4 Part A: Participatory Observation

3.4.1 Methods

3.4.1.1 Participant Selection

Participatory observation was undertaken to better understand the active creation and development of VGI with members of the OpenStreetMap user group; intended as a snapshot insight into the culture and perspectives. Because the Google Maps and Ordnance Survey map projects focus on the use of neogeography rather than the creation of information from volunteer sources, they were not investigated in such a way. Events with which to participate and observe within were found in the following ways:

- OpenStreetMap mapping parties within 100 km of Loughborough were discovered through the OSM events calendar (http://wiki.openstreetmap.org/wiki/Current_events).

- Contact with the OpenStreetMap community via the news section of the website.
- Personal contacts within Ordnance Survey.

3.4.1.2 Observation Design

As McCall and Simmons (1969) noted, participatory observation *involves repeated, genuine social interaction on the scene with the subjects themselves as part of the data-gathering process*. Within this study, the position of *marginal participant* was sought (Gold 1969; Junker 1960) to allow a higher degree of involvement and insight than the passive position of observer-as-participant; yet without the high involvement of participant-as-observer. Participatory observation took the form of attending various OpenStreetMap mapping parties to generate VGI data within the community and attending VGI and PGI focused conferences to talk with users.

3.4.1.3 Procedure

Data for participatory observation was captured using *descriptive observation* for the various scenarios of focus; see Table 3.1. Rather than take notes during observation, events were recorded after participatory observation has taken place, allowing for greater emersion within the activities.

3.4.1.4 Analysis

McCall and Simmons (1969) stated that the output from participatory observation is *an analytic description of a complex social organisation*. This resulted in three key elements of the analytical description:

Table 3.1 Dimensions of descriptive observation (Spradley 1980)

Descriptor	Definition
Space	Layout of the physical setting; room, outdoor spaces, etc.
Actors	The names and relevant details of the people involved
Activities	The various activities of the actors
Objects	Physical elements: furniture, etc.
Acts	Specific individual actions
Events	Particular occasions, e.g. meetings
Time	The sequence of events
Goals	What actors are attempting to accomplish
Feelings	Emotions in particular contexts

3.4 Part A: Participatory Observation

1. Employing concepts, proposition and empirical generalisations of a body of scientific theory as the basic guides in analysis and reporting
2. Thorough and systematic collection, classification and reporting of facts
3. Generating new empirical generalisations.

Records of observation were not coded, yet the statements and outcomes were used as an alternative perspective on the outcomes from the focus groups.

3.4.2 Results and Analysis

Participatory observation occurred on four occasions, involving over 50 different users of VGI associated with OpenStreetMap. In addition to taking part in the data collection and mapping session, the OpenStreetMap 'State of the Map' conference was also attended; see Fig. 3.1. This gave insight into the thoughts, feelings and actions of OSM members of the course of a few days in both formal and informal environments. Topics covered during this time included data collection, social interaction, contribution, perspectives on other map platforms and the meaning OSM has to the contributors on a personal level.

The following key outcomes were derived from the participatory observation during the study:

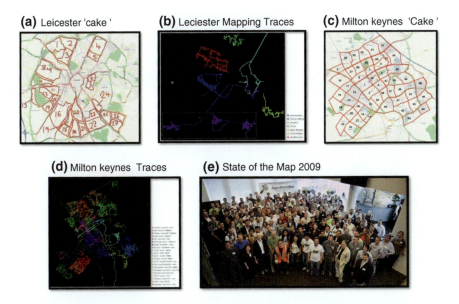

Fig. 3.1 Examples of participatory observation. **a** Illife (2009), **b** Sward (2009), **c** Wood (2009), **d** O'Brien (2009), **e** Gil Biraud (2009).

- Social interaction is a central activity at VGI data collection events, particularly comradery at the joint effort of creating *the map* which they feel will help influence society at large for the better.
  - *Example: Strong social interaction before and after mapping parties, more weekly meetings of members in pubs than mapping parties.*
- Anarchic organisation; i.e. participants chose to engage with events due to their personal interest in their application rather than because of a prerequisite.
  - *Example: Mapping parties organised by anyone for any reason without the need to comply to any guidelines or practices. Many mappers contributing vast amounts of data without engaging in the social functions, online discussions or other forms or guidance.*
- Celebration of achievements, yet not much recognition of gaps in the data; e.g. celebrating mapping one section of the city as a triumph, yet ignoring the other sections still blank and not surveyed.
  - *Example: The State Of The Map conference focused heavily on achievements and developments within the community, without recognition of the pitfalls, shortcomings or errors within the data set (as mentioned by participants within interviews).*
- Optimistic and exciting outlook driven by potential of the map rather than its current form.
  - *Example: Mapping party participants talked with much enthusiasm about what the map will be like, how it will be used and future developments before and after mapping sessions.*
- Hostile towards criticism, especially from those outside of their group, even when giving a balanced appraisal.
  - *Example: Non-regular mappers at the mapping parties who voiced concern over validity or completeness were not brought into much discussion and non-verbally 'shut out' by some members.*
- Low levels of standardisation towards how data should be captured, contributed and utilised. In particular, each instance of observation had a different outlook on these matters.
  - *Example: Much discussion at the mapping parties on how data could be captured, contributed and edited, without a single voice of universal agreement.*
- Post data collection, there was limited feedback from the organisers on achievements or continued engagement with members of the group.
  - *Example: No procedure of follow up emails or forums posts following any of the user engagements.*

- Keen interest in geography in general from the participants, choosing to refer to more technical terms over standard terms wherever possible.
  - *Example: Specifying the meeting point (Leicester University canteen) not by name or address, but by GPS coordinates.*

During these sessions, there was no evidence of users consulting professional information sources to confirm locations or features in the built environment. Interestingly, this extended to locating the meeting point, where GPS coordinates were given rather than an address in order to *add to the spirit* of the occasion. While non-referral to PGI sources while actually mapping may be taken as essential to avoid copyright infringement, other such extreme measures demonstrated the strong sense of independence within the VGI community.

3.5 Part B: Interviews

3.5.1 Methods

3.5.1.1 Participant Selection

At the start of this book, examples of research which demonstrated a useful categorisation of users associated with VGI were limited. However, in a study investigating the data quality issues within VGI, Coote and Rackham (2008) grouped users into four categories: *consumers, special interest [mapping] groups, local communities* and *professionals*; see Table 3.2. Although these users may not be mutually exclusive, (i.e. a user may be only a *consumer*, or also a *consumer* and a *producer* of VGI) this simplified model offered an effective framework of exploration.[1]

Consequently, participants were recruited in each of the four categories. Additionally, within each of the categories a range of participants was sought who represented at least one of the three main map products (see Sect. 3.3.1, page 24). Finally, the participants were required to fit the following specification:

- Regular involvement with their map product;
- Use of the map product for work or social reasons involving relating information to geographic locations;
- Have awareness of map products outside their chosen product;
- Aged 18–65, being a non-vulnerable person according to the Loughborough Ethics Guidelines.

---

[1] Following the completion of this study this distinction between the various users were highlighted as being useful for understanding the user interactions of VGI and offers a potentially beneficial framework for human factors investigation (Brando and Bucher 2010; Brown et al. 2012).

Table 3.2 Segmentation of target respondent user groups (Coote and Rackham 2008)

User Group	Characteristics
Consumers	A person who purchases [or selects] any product or service for personal use
Special Interest Groups (*SIG*)	Individuals who come together to collaboratively achieve some shared goal
Local Communities (*LC*)	local people who have a common desire to improve their local area
Professionals	Users employed by organisations that use geographic data to perform their business activities, whether to analyse, report, navigate or otherwise maintain systems

In order to find participants who fitted the above criteria, the following recruitment techniques were employed:

- OpenStreetMap mapping parties were attended (Leicester and Milton Keynes) where contacts were made, flyers handed out and users discovered.[2]
- Posters advertising for Google My-Maps users to take part in this study were placed around Loughborough University student areas (e.g. Student Union, departments and halls of residence).
- OpenStreetMap State of the Map 2009 conference was attended where contacts were made, flyers handed out and users discovered.
- Searches for keywords such as *Google Maps* and *My-Maps* were conducted on Twitter, with results refined to the local areas (e.g. search for *'My-Maps' Google near:nottingham*).
- Email adverts for participation in the study were posted on the OpenStreetMap Mailing lists for the UK.

3.5.1.2 Theoretical Justification

Lin et al. (2005) commented that two key measures of value exist; *unidimensional* (measuring customers overall perception of value) and *multidimensional* (measuring the various value perceptions using various benefit and sacrifice dimensions) perspectives. As noted by Sheth et al. (1991), both have been demonstrated as being useful in understanding (and predicting) user behaviour.

The unidimensional theory of value can be seen as the benefits and sacrifices associated with only one element of perceived value, e.g. price or service (Lin et al. 2005). However, Sweeney and Soutar (2001) noted *a more sophisticated measure is needed to understand how consumers value products and services.* Further to this Lin et al. (2005) noted that the *unidimensional conceptualization strategy is effective and straightforward, but it cannot discern the complex nature of perceived value.*

---

[2] **Mapping Party:** VGI Contributors to the OpenStreetMap project getting together to do some mapping, socialising and chat about making a free map of the world (OpenStreetMap 2011).

3.5 Part B: Interviews

In defining the multidimensional perspective, Sweeney and Soutar (2001) included the components of *emotion, social enhancement, price* and *performance*. Within this model, each construct may be considered a *give*, a *get,* or a considered trade-off between the two. Crucially, the multidimensional perspective considers all of the various value dimensions together, rather than the independent factors under the unidimensional perspective.

Importantly, the dimensions within the theory of multidimensional value are not fixed, as shown by the various contentions by authors such as Sheth et al. (1991) and Zeitham (1988). Therefore, two conclusions may be drawn:

1. Due to the currently unknown, yet assumed complex nature of neogeography, the most appropriate theory of value to be used within this study was the multidimensional theory.
2. The dimensions which best predict the value perceptions of the uses are currently unknown. Therefore, the theoretical framework should start with the basic elements of *emotion, social enhancement, price* and *performance*, yet be prepared to adjust for the dimensions emerging from data analysis.

3.5.1.3 Interview Design

In order to extract the most relevant information from the participants during the interview, it was necessary to base the questions posed on the theoretical framework that would be used to analyse the transcripts.

From an interaction design perspective to help understand the reasoning and expression of the themes and effects of user relationships in system design, Monk and Howard (1998) developed the tool of the rich picture. Development is attained through analysing transcripts for references to other users, communication and data flow between users, as well as tensions and concerns of all those involved. It was the intention that the representation of user interactions would provide a framework to contextualise outcomes from the interview.

In order for the interviews to produce adequate results by which a rich picture may be drawn to represent inter-user relationships (Monk and Howard 1998) and multidimensional value perceptions inferred, the interview question sheet was split into two sections, each addressing a different research requirement; see Table 3.3.

Table 3.3 Sections and themes required of the interview question sheet

Section	Investigation themes
Rich Picture—Ecology of the User (Monk and Howard 1998)	Connections
	Tensions
	Data transfer
	Knowledge of other parties
Perceptions of VG—Multidimensional Value (Sweeney and Soutar 2001)	Emotion
	Social enhancement
	Price
	Performance

Although questions were designed to focus on each of the components highlighted in Table 3.3, they were open ended enough for the participant to discuss whichever topics or themes they felt more relevant to them. Consequently, the following categories of question were employed:

1. Involvement in mapping project;
2. Background relative to project involvement;
3. Influence of project on life;
4. Interaction with others;
5. Feelings of completeness in mapping project;
6. Feelings towards user-generated content;
7. Missing features within mapping project;
8. Contribution of information in general;
9. Application of mapping project.

3.5.1.4 Procedure

Participants were contacted through email, personal communication and internet forums (e.g. forum.openstreetmap.org). Interviews were arranged for semi-public locations (e.g. coffee shops, libraries, etc.) at a time and place to suit the participant. Before the interview, full information as to the purpose the interview and how the data would be used was presented to the participant before consent being obtained. During the interview, an audio recording was taken to capture all questions and responses in detail, in line with established practice with interviews and analysis (Lapadat and Lindsay 1998). Main questions were asked, with supplementary probing questions following to fully explore the topic areas.

3.5.1.5 Data Analysis

Interviews were recorded and later transcribed in full. In order to produce a deeper insight into the dimensions of value within the transcripts, thematic analysis was conducted. It was important to understand how those value dimensions described the user's perception of VGI (e.g. how do users feel emotionally about the subject). Consequently, each value dimension was considered from a *gains and sacrifices* perspective, similar to that offered by the unidimensional theory of value (Lin et al. 2005).

3.5.2 *Results and Analysis*

Over the course of the study, 16 participants were interviewed. Qualitative analysis of the semi-structured interviews centred on understanding the relationships

3.5 Part B: Interviews

between user groups. In particular, describing the similarities and differences in how they operate and perceive both VGI and PGI. Importantly, these qualitative outcomes were used within this book as guidelines to user perceptions rather than as developed theory.

3.5.2.1 User Relationships

Below is the rich picture developed through qualitative analysis of the transcripts and participatory observation, demonstration data flow (arrows), concerns (thought bubbles), and tensions (swords) between the various users associated with neogeography.

Table 3.4 presents a key to the features used in the rich picture:

3.5.2.2 Inter-User Data Flow

The most basic flow of data are from the producers (i.e. professionals, Special Interest Groups and local communities) to the consumer; i.e. the end user. The consumer does not return data to any sources as doing so would make them a contributor. The exception to this case could be where data are contributed to a mapping project unintentionally, as with the example of the *Tom-Tom HD Traffic Service* (Palmer 2008).

Within groups, the data flow is relative to the structure of the organisation. For example, within traditional mapping agencies, flow of data relating to GI follows a managed, intentional and structured path from generation through to quality control and distribution. Within SIGs, data are shared openly amongst all members, with free expression of views and equal opportunity in development. The internal flow of data in both organisations is little observed and to an extent has little influence of those utilising their product; the maps they are generating. In professional organisations, trying to find a business model that would enable current data integrity while utilising the potential of VGI causes some tension as to the future direction of the company.

Within SIGs (being loose organisations with less structure than a formal corporation is) the main form of communication is through Wiki's and mailing lists. Although working as an effective form of communication for levelling and

Table 3.4 Key to symbols used in Fig. 3.2 (Monk and Howard 1998)	Symbol	Meaning
	Crossed swords	Tension between user groups
	Arrows	Data flow (in direction of arrow)
	Thought bubbles	Concerns of users
	Cartoon icons	User groups
	Yellow boxes	Project groups

democratising an organisation, these are the main channels of tension within these groups, causing on-going *back and forth* mailing list arguments known as *flame wars*:

> It's an interesting time for OpenStreetMap and CloudMade as well as you can see some quite aggressive comments going back and forwards about, people now turning what they thought was a community project into a professional service; information services [#1-12]

Professionals receive data (VGI and/or PGI) from the producers in a similar way that consumers do, yet with greater access to data sets or technical capacity. This allows greater exploitation and customisation of their licensed map. Tensions arise when the cost of the data from proprietary producers is too high for their business model, causing lower return on investment than desirable, or when VGI is not up to their desired specification.

3.5.3 Multidimensional Value Dimensions

This section presents a breakdown of the multidimensional perspectives of value relative to the user groups investigated through this research. Through thematic analysis it was discovered that the dimensions of *emotion, price, performance, social, epistemic* and *conditional* were useful categories for describing user value (Sheth et al. 1991; Sweeney and Soutar 2001). However, categories of *legal* and *moral* dimensions were observed and are therefore included.

3.5.3.1 Emotional Value

VGI contributors have an emotional connection to subject

> I don't do OpenStreetMap because I feel I have to; I do it because I get a warm fuzzy feeling out of doing it [#1-02]

The strong emotional attachment of the contributors (not demonstrated by consumers) is the reason for their continued involvement in the VGI project. This may be seeing the continued improvement of their product or their contributions, and therefore is less likely to influence the consumers of any VGI products.

Users (not PGI professionals) are concerned about data vandalism

> One thing people always worry about is vandalism, people intentionally putting in… erroneous data [#1-07]

The emotional concern of the users towards the data accuracy is not ideological, but revolves around the trust placed in the contributing community to deliver information which is accurate and reliable every time.

3.5.3.2 Functional Value

VGI presents the zeitgeist of contributor interest

> You start to discover areas that have only just been built, new shopping malls for example, and you also come across social geography as well. So we don't only look at the spatial coordinates associated with photos, we also look at the tags which are associated with it [#1-11]

This enabled companies with a geographic interest to make use of VGI in a new way. For example, if a region receives many contributions it reflects high activity geographically, and the data they contribute indicates the areas of interest.

Users from all groups feel that their neogeographic project is better than the competition

> If I was completely abstract from OpenStreetMap, you'd look at OpenStreetMap and you would see there is more information, there is more things that you can look at. I mean, you just look at the centres of Amsterdam; it's even marked the prostitution areas [#1-02]

This experience may be explained by the users utilising one map over another for a personal (and potentially unique) reason. Additionally, prior preferences and bias may provide a key element of product choice.

VGI enables information not found on traditional maps to be utilised

> The practical side of it is there is no other system available that can give me the bits of maps that I want, like only maps with footpaths, and with bicycle parking, and with bike shops, and with this that and the other [#1-03]

Non-commercial niche mapping may be one of the greatest strengths of VGI from the consumers' perspective, providing a specific product they want rather than a *generic* map. A good proportion of SIGs and consumers desire more local information presented and accessible from their chosen map. This suggests that extra information not found on traditional maps may be a very important part of the user perception of VGI.

Users perceived VGI as accurate enough for their needs

> Giving me routes from one place to another... it doesn't actually need it to be perfect for it to still allow me to do what I need to do [#1-03]

The arguments against VGI use based on its accuracy may be correct, but not relevant from the user's perspective. The strength at which this functional perception is supported across project groups is contradicted by the number of users who perceive VGI as *not completely trustworthy*.

Users cannot always trust VGI

> You can't trust it 100 % at any one time, especially because you have no idea who just messed it up last week, but nobody's noticed yet [#1-03]

This mixture of opinions over how much trust may be placed in VGI suggests some bias in the user base, e.g. they do not feel they can fully trust it, yet in practice, they can.

Users see mapping in regions not covered by PGI as a strong benefit of VGI

> *There's a person… who's working in Gaza at the moment making maps of Gaza which are being used by aid agencies… Now these are cities that… have no official maps, because the roads have been coursed, they haven't been planned, they've just occurred [#1-02]*

Although this may benefit travellers to developing countries, this is unlikely to directly impact the general public within (for example) the UK who perceive this as a benefit.

SIGs associated consider VGI to be more up to date than PGI

> *You've got the physical route that is essentially the most current. I mean I've been going in Wales on the Crib Loch path to Snowdon and the problem was that the path had changed. On the Ordnance Survey map it said it went 'this way' around the 'pig path', when in reality it went the other way [#1-02]*

This highlights one of the potential strengths of VGI, how changes in human activities may be recorded and reflected with VGI to a much higher degree than through traditional cartography. However, this perception is not shared with any of the interviewed participants outside of the OpenStreetMap project, and may be related to their involvement in development of the base map.

The ability to customise or personalise maps with VGI is of benefit to work

> *I'm a member of the cyclist touring club… so mapping is essential for that, and when you come to cycle campaigning, working out cycle routes again involves mapping [#1-05]*

This benefit may be associated with neogeography, delivering the ability to collaboratively work on a single project from remote locations with few time or technology limitations. However, within this study it is the VGI contributions which are powering such benefits since the information they use cannot come from PGI sources.

3.5.3.3 Knowledge Value

VGI provides an increase in local knowledge from mapping their own area

> *It's also an occasional excuse when I can get off my backside and to go and explore parts of Leicester that I really think I would rather not know about [#1-01]*

This benefit is potentially an important motivation factor for continued contribution to VGI projects. However, it does not affect users outside of SIGs, unless this benefit is used to help recruit consumers to become contributors.

3.5.3.4 Legal Value

Users enjoy freedom to *do what they like* with the map data

> *I use OpenStreetMap as my data set because it's a free and open-source version of the dataset. I don't have to pay a Navteq or Google for their data and also its relatively adjustable, which for myself as a student and as an entrepreneur, I can take that data set and do anything I want with it without cost considerations [#1-02]*

From a business perspective the *do what I want* mentality removes barriers to innovation so that full utilisation is possible. However, this applies only to the open source examples of VGI (e.g. OpenStreetMap) but not closed source VGI; e.g. Google Map Maker.

3.5.3.5 Moral Value

VGI benefits others

> *I also like the idea of helping someone in an area that's not going to get the love of the companies because it just isn't viable for them. Whereas you can help someone because you want to. [#1-08]*

This perception was particularly strong in the SIG category, possibly due to their direct involvement in VGI for other [potentially anonymous] users. This altruism may be a motivating factor for contributors to continue contributing, or to help recruit consumers into becoming contributors. Professionals may use such VGI may increase their company or product image.

Open source VGI fits the ideology of contributors

> *There's an ideological drive behind it as well. Behind the licensing, this is where it cuts different from just being a great map, is the license allows you to do things with it, gives you almost unrestricted access to whatever creative thing you come along with and so in the same way it doesn't matter how little the cost of software is, the free software, the open source software is still important to me, and it's the same with the mapping stuff [#1-03]*

This suggests a difference in the outlook between contributors and professionals, potentially a barrier to cross collaboration, such as SIGs not wanting to contribute to a professional/proprietary project on ideological grounds.

3.5.3.6 Price Value

The zero cost to access VGI maps is a large benefit to the interest of SIGs

> *Is that not part of what the whole thing's about, so people can generate maps for themselves without having to pay extortionate amounts? [#1-04]*

This perception may be relative to the legal perspectives of open licences allowing users to *do what they like*. The importance of this may also be seen relative to Rogers (2003) perspective that the zero price tag opens up the ability for the user to try the product out, and therefore helps increase the utilisation of the innovation in the community. Capitalising on this perception from a consumer's perspective may help to increase use and overall positive judgements of VGI.

3.5.3.7 Social Value

An enjoyable community of VGI contributors and developers

> *Before, I was one of a number of contributors and I was able to actively actually develop for OpenStreetMap when I was working in Cloud Made. Also the access you get, I mean by sitting over a pint or a coffee and just explain, talking to the founder, he explained his motivations, and then you see the internal workings etc. [#1-02]*

It is possible that this strong community bond within these groups increases the overall perception of value of VGI (Sweeney and Soutar 2001), keeping users involved with VGI. However possibly consumers and PGI professionals did not express any benefits of community involvement, outside their own workplace or organisation; separate from all GI.

Collecting and contributing VGI takes up personal time

> *Sacrifices... its time that'd be spent doing other things. My shed has needed reroofing since the middle of winter when the frost got to the felt, and I've still not got round to doing it... I'd rather be mapping than working hard on the shed [#1-04]*

The investment of time to learn the skills and to actively partake in geographic contributions could be a barrier to some users becoming involved in a VGI project. Alternatively it may cause slow progress or participants ending their contributions. The sacrifice of the free time of users to contribute VGI may act as a barrier to users becoming contributors.

3.5.4 General User Perspectives

The following is a summary of characterises for user groups within this study, intended to relate to user interactions represented in the rich picture.

3.5.4.1 Map Product Use

Consumers **select their map to fit their circumstances with little loyalty**

> *Apart from using it like everybody does in terms of looking for places and directions, I've used Google My Maps, at the moment mainly for my own use... I've used it in a work context because I was trying to organise a meeting [#1-10]*

Consumers may be open to using (or at least trying) new map products from both VGI and PGI sources. However, emphasis needs to be placed on the utility and usability of such products rather than to expect product use based on the authority of the contributor.

***Special Interest Mapping Groups* (SIMGs), *Special Interest Mapping Group Contributors* (SIMGCs) and *Professionals* are loyal in the use of their group's map**

> I'll often check out to see if the local CTC has a website [same map project involved in] to see what's on there. And being able to find where the tea places are in the locality is quite useful [#1-05]

Observation of SIG members also showed a great bias towards their map product (i.e. OpenStreetMap) and hostility towards rival map product. This was often in spite of rival map products with opinions that were in some cases unfounded. This may limit the ability for cross-collaboration between map projects based on the low desire to switch to a different product.

3.5.4.2 Information Use

***SIMGCs* produce data for group members and external parties to use their data**

> It's mainly just a project to collect data... we hope other people will use it for whatever they feel free to use it for [#1-08]

While contributors may also be consumers, they product VGI for the sake of its production rather than for specific pre-determined tasks with known outputs.

***Professionals* however take the VGI combine it with VGI as long as it enhances their business position**

> The major proprietary vendors operate within the PND market sector, so Personal Navigation Device. If you can drive to it, great. If you can drive to it in an area of the world where the economy is sufficient to support a burgeoning Sat-Nav and hand-held community, great. Outside of that data uptake and data penetration is marginal; it's very slow. And that means areas of the world are basically blank, and OpenStreetMap enables those blank areas to be filled in [#1-11]

This shows a real benefit for VGI to be used alongside PGI in mashups and consumer products, but it relies on the VGI meeting strict requirements and the demonstration that it will enhance the user judgements of the product.

3.5.4.3 Accuracy

***SIMGCs* are less concerned about inaccuracies in data than *consumers* are as they have a stake in improving the data**

> It has its faults but there are no glaring errors... It's very much if you don't like it you can fix it yourself which appeals to my, well, sense of working I suppose [#1-02]

The perspective which SIMGCs have for the data may be out of sync with the feelings of the consumers. Therefore, better ways of filtering the data, or quality control should be implemented which meet the needs and concerns of the consumers.

Professionals **are concerned about data validity, how inaccuracies may hurt their business position and show concern over what VGI actually means to their customers**

> *If I'm dispatching ambulances, and I know that I need to get to the patient within 7 min, can I trust the volunteer captured information? [#1-12]*

Although VGI has potential to be fully incorporated into the business plan of companies, a way of measuring quality assurance, or guaranteeing the accuracy and currency of the VGI is required.

3.5.4.4 Influence on VGI

Those not involved in the contribution and development of VGI have little influence on the product

> *All we can do is we can influence the direction this takes by offering suggestions [#1-11]*

Low influence may cause a lack of understanding from the VGI producers as to what the consumers need and want. Consequently, they risk producing highly interesting products with limited consumer utility.

3.6 Discussion

3.6.1 User Value Dimensions

Participants generally perceived that the *quantity* and *salience* of benefits outweighed the sacrifices involved in the use of VGI. Consequently, the participants tended to judge VGI as a product of high personal value when considering their overall appraisal of the subject matter. Table 3.5 presents the breakdown of value dimensions drawn from the thematic analysis of the interviews, in relation to those value elements mentioned as being important to the user within the literature.

Table 3.5 shows how *moral* (the user's basis of what is right and wrong) and *legal* (items relative to positions of statue in the law) values appear as salient categories of user judgements (Bruns 2008; Coleman et al. 2009). However, these dimensions are not included within the multidimensional value theories of value; the theoretical framework of this study.

Within a consumer purchase context, Carrigan and Attalla (2001) remarked that most consumers pay little heed to ethical considerations in their purchase

3.6 Discussion

Table 3.5 Analysis of value dimensions used in the scoping study

Value element	Suggested in literature (Sweeney and Soutar 2001)	Suggested in literature (Sheth et al. 1991)	Emerged from the scoping study
Emotion	Yes	Yes	Yes
Price	Yes	–	Yes
Functional	Yes	Yes	Yes
Social	Yes	Yes	Yes
Epistemic	–	Yes	Yes
Conditional	–	Yes	–
Legal	–	–	Yes
Moral	–	–	Yes

decision-making behaviours. However, the strongest response to moral value as a construct of their multidimensional value in VGI was from SIMGs. This is possibly a result of their ideology in contributing to a wider community without an obvious personal return for their efforts. Such extrinsic perspectives in open source contributors was highlighted by Lakhani and (2003) in that open source contributors participate in these projects in part to help and aid others. This—in relation to the work of Carrigan and Attalla (2001)—demonstrates moral value as an important value dimension, particularly with *SIMGs*.

Legal issues were seen as salient amongst VGI related users with emphasis on the freedom to manipulate and use data without restrictions. In support of this, comments made by VGI contributors during participatory observation were generally hostile towards legal limitations on data access. This may be considered a constant undertone, explaining why the legal dimensions came through in the thematic analysis.

Currently the legal dimension is not discussed inside consumer activity related value theory. However, Coleman (2009) highlighted that within the open source community such freedoms are seen as intrinsic to the liberal freedom of expression and human rights relating to technical ability. While Lakhani and Wolf (2003) demonstrated the personal reasons for contribution, the legal freedoms allowed to the user through the open source licences facilitate these activities. This made the personal enjoyment, fulfilment, challenge and social enhancement possible. Consequently, the salience of legal value within a multidimensional context is useful in highlighting the attitudes of those associated with VGI more than describing the practices of VGI.

Table 3.5 shows the themes that emerged from the coding of this study contain a stronger correlation with the work of Sweeny and Soutar (2001) than the work of Sheth et al. (1991). Additionally, the categories used include *price* and exclude *conditional knowledge*, making the work or Sheth et al. less relevant. However, correlation with other dimensions within the results is still relatively high. This affords an additional richness in describing the user perceptions and reactions to VGI within neogeography.

3.6.2 Spatial-Data Infrastructure (SDI) Relationships

The rich picture presented a complex and dynamic series of relationships between the users associated with VGI. This study has demonstrated that as a tool for understanding such complexities, the rich picture is useful in creating an easily accessible framework to which further findings can be contextualised. Therefore, consideration should be given to the similarities between the rich picture and user relationships presented through the literature.

Budhathoki et al. (2008) and Grira et al. (2010) presented a framework where all users associated with VGI communicate with each other. Here the strongest connections exist between *expert organisational users* and *expert organisational producers*. One of the key contributions that this study has made is to place boundaries on this notion. The rich picture suggests that while—within a given community (e.g. SIGs such as OpenStreetMap)—the infrastructure as described by Budhathoki et al. (2008) may hold true, the model of Budhathoki et al. does not describe the full range of users associated with VGI.

An additional insight into the relevance of the rich picture as a tool can be seen through the *participatory observation* and *interviews*. This showed that the tensions between user groups is affected by the user group ideology (and thus in part the clashes between user group ideologies) and the form of data the users interact with. This in turn affects the flow of information within the wider user group infrastructure described by Fig. 3.2, p. 27. During participatory observation, such tensions were observed in how just mentioning proprietary data to VGI contributors provoked highly hostile and negative comments, while affirming the virtues of their own projects. Such perspectives are not covered by the simplistic model offered by Budhathoki et al. (2008). Therefore, this study has found that the more complex and insightful rich picture of Fig. 3.2 to be useful in understanding and relating the experiences and information judgements of users.

3.7 Conclusions

3.7.1 Relating to the Project Aims

To assess the success of this study in addressing the research aims of this book, consideration should first be given to how successfully the study aims have been addressed.

1. *What Is The Nature Of VGI*

    This study has also shown how VGI (such as OpenStreetMap) is predominantly being produced by members of *Special Interest Groups*, who also develop the VGI systems as a community for utilisation by *Professionals* and delivery to

3.7 Conclusions

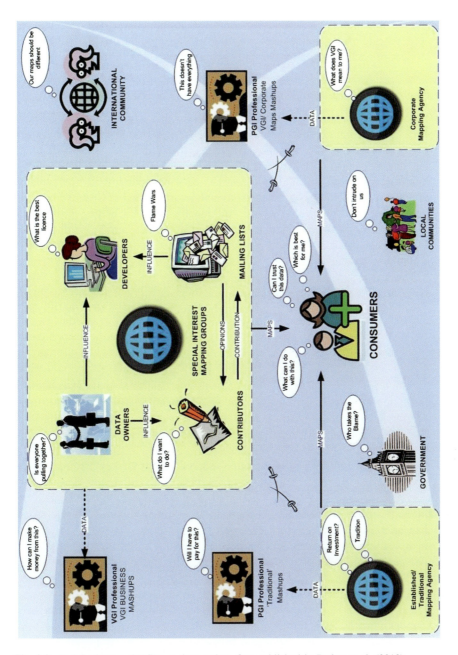

Fig. 3.2 A rich picture of VGI user interaction; first published in Parker et al. (2010)

consumers. The majority of information flow in this context is between users occurs inside project groups (e.g. OpenStreetMap, Google My-Maps), with the product of each group being the inter project/group data for transfer.

2. *What Are The Different Characteristics Of The Key Users*

The main outcome from this research has been that while users of VGI may often share common perceptions (e.g. *SIMGs, SIMGCs* and *professionals* having a vested in the use of their groups' map), different users will often perceive elements of VGI differently, based on which user group they may be identified with and the VGI project they are interacting with. The greater outcome of this study has been the examination of how and to what extent these similarities and differences occur. Additionally, the rich picture provided a visual framework to identify the interaction of users in terms of information flow between users; and inter-group tensions relative to those users investigated in this scoping study.

Through participant interviews and participatory observation, one prevalent theme has been that those users who are involved in VGI (OpenStreetMap) contribution and development are more biased towards their VGI project, and more against PGI projects than non-involved users may be.

3. *Understand How Different Users Perceive VGI*

Although this study was based upon value theory, determining a user-collective perception of value is an elusive concept (Zeithaml 1988). However, if considering value as the improvement to a users' condition through utilising VGI (Menou 1995), then a salient increase in user value can be observed in all functional and work related perceptions.

The analysis of user perceived value supports one of the key assumptions of this book, that different groups of users perceive VGI differently. This is possibly due to each user group having its own needs and objectives causing different aspects of the same phenomenon to be more important to one group than another. The relation of user perceptions within the multidimensional theory of value have been demonstrated as relevant to the assessment of VGI user perceptions. However moral constructs were perceived as salient within the SIMG user group despite not being mentioned as an important in user perception in traditional value theory (Carrigan and Attalla 2001; Sweeney and Soutar 2001). Due to the *emotional, moral* and *social* salience of user perceptions towards VGI, the theories of *Worth Centred Design* and *Value Sensitive Design* are highly applicable to the activity of designing applications which utilise VGI; especially relating to SIMGs. However, *Value Centred Design* may not be as applicable to such VGI projects.

3.7.2 Relating to the Research Questions

This scoping study provided a useful insight into the how VGI is generated and utilised within a variety of situations by a complex network of users. However, this study has also identified that the conditions of information generation and utilisation required by *Research Question One* are relative to the nature of the VGI project, the reason the user is accessing the information relative to task and its unique user community. Consequently, while generalisations may be drawn on these factors, future investigation into VGI from a *User Centred Design* perspective must treat each VGI project as unique in its own right to best design for its users.

This study also highlighted how while the nature of VGI and PGI may at times be similar, the ways in which these two information forms are processed by professions and utilised by consumers can provide a clear distinction. In addressing *Research Question Two*, this study led to the production of a clear framework of VGI. This demonstrates these similarities and differences, as well as providing a framework for understanding the Neogeographic phenomenon.

Ultimately, the scoping study serves as a useful framework to contextualise the way in which different users perceive VGI. In order to successfully build on these outcomes, an understanding as to the ways consumers utilise and perceive VGI in relation to PGI is required. In particular, it is essential that further investigation focuses on existing use of information by consumers to produce useful outcomes within a design context.

References

Harding J et al (2009) Usable geographic information—what does it mean to users? In: proceedings of the AGI GeoCommunity'09 conference, AGI GeoCommunity, Stratford-Upon-Avon, UK

Brown M et al (2012) Usability of geographic information; current challenges and future directions. Appl Ergonomics 44(6):855–865. Available at: http://www.sciencedirect.com/science/article/pii/S0003687012001718

Brando C, Bucher B (2010) Quality in user generated spatial content: a matter of specifications. In: 13th AGILE international conference on geographic information science, AGILE, Guimarães

Bruns A (2008) Blogs, Wikipedia, Second Life, and Beyond. Peter Lang Publishing Inc, New York

Budhathoki NR, Bruce B (Chip), Nedovic-Budic Z (2008) Reconceptualizing the role of the user of spatial data infrastructure. GeoJ 72(3):149–160. Available at: http://link.springer.com/article/10.1007%2Fs10708-008-9189-x?LI=true#page-2

Carrigan M, Attalla A (2001) The myth of the ethical consumer—do ethics matter in purchase behaviour? J Consum Mark 18(7):560–577

Coleman G (2009) Code is speech: legal tinkering, expertise, and protest among free and open source software developers. Cult Anthropol 24(3):420–454

Coleman DJ, Georgiadou Y, Labonte J (2009) Volunteered geographic information: the nature and motivation of produsers. Int J Spat Data Infrastruct Res 4:332–358

Coote A, Rackham L (2008) Neogeography data quality—is it an issue? In: Holcroft C (ed). Proceedings of AGI Geocommunity'08, Association for Geographic Information (AGI), Stratford-Upon-Avon, UK, p 1. Available at: http://www.agi.org.uk/SITE/UPLOAD/DOCUMENT/Events/AGI2008/Papers/AndyCoote.pdf

Foth M et al (2009) The second life of urban planning? using neogeography tools for community engagement. J Location Based Serv 3(2):97–117

Gil Biraud ME (2009) SOTM09 group photo.jpg. Available at: http://www.flickr.com/photos/mgilbir/3715653401/in/pool-sotm09

Gold RL (1969) Roles in sociological field observation. In: McCall GJ, Simmons JL (eds) Issues in participant observation: a text and reader. Addison-Wesley, USA, pp 30–38

Goodchild MF (2007a) Citizens as sensors: the world of volunteered geography. GeoJournal 69(4):211–221. Available at: http://www.springerlink.com/content/h013jk125081j628/

Google (2010) Beyond the basics: my maps. maps.google.com, 2010 (January 4th). Available at: http://maps.google.com/support/bin/answer.py?hl=en&answer=68480. Accessed 3 April 2012

Grira J, Bédard Y, Roche S (2010) Spatial data uncertainty in the VGI world: going from consumer to producer. Geomatica 64(1):61–72

Haklay M (2010a) How good is volunteered geographical information? a comparative study of openstreetmap and ordnance survey datasets. Environ Plann B 37(4):682–703

Haklay M, Singleton A, Parker C (2008) Web mapping 2.0: the neogeography of the geoweb. Geogr Compass 2(6):2011–2039

Illife M (2009) Leicestercake01

Junker B (1960) Field work. University of Chicago Press, Chicago

Lakhani KR, Wolf RG (2003) Why hackers do what they do: understanding motivation and effort in free/open source software projects., working pa

Lapadat JC, Lindsay CA (1998) Examining transcription: a theory laden methodology. In: annual meeting of the American educational research association. American Educational Research Association, San Diego, USA, p 3

Lin C-H, Sher PJ, Shih H-Y (2005) Past progress and future directions in conceptualizing customer perceived value. Int J Serv Ind Manage 16(4):318–336. Available at: http://www.ingentaconnect.com/content/mcb/085/2005/00000016/00000004/art00001

McCall GJ, Simmons JL (1969) Issues in participant observation: a text and reader. Addison-Wesley, USA

Menou MJ (1995) The impact of information II: concepts of information and its value. Inf Process Manage 31(4):479–490

Miller CC (2006) A beast in the field: the google maps mashup as GIS/2. Cartographica: Int J Geogr Inf Geovisualization 41(3):187–199

Monk A, Howard S (1998) The rich picture: a tool for reasoning about work context. Interact 5(2):21–30

O'Brien O (2009) MK party render.png

OpenStreetMap (2009) Beginners' Guide. wiki.openstreetmap.org, 2010 (January 4th). Available at: http://wiki.openstreetmap.org/wiki/Beginners'_Guide. Accessed 4 Jan 2010

OpenStreetMap (2011) Mapping parties. wiki. openstreetmap.org, 2011 (July 23rd). Available at: http://wiki.openstreetmap.org/wiki/Mapping_party. Accessed 23 July 2011

Palmer J (2008) Tech that trumps traffic tangles. BBC News, 2010 (January 12th). Available at: http://news.bbc.co.uk/1/hi/sci/tech/7733919.stm. Accessed 2 April 2013

Parker CJ, May AJ, Mitchell V (2010) Characteristics of VGI stakeholders, 2010 (November 29th). Available at: http://www.slideshare.net/kyral210/characteristics-of-vgi-stakeholders-3536826

Rinner C, Kumari J, Mavedati S (2011) A geospatial web application to map observations and opinions in environmental planning. In: Li S, Dragicevic S, Veenendaal B (eds) Advances in WebGIS, mapping services and applications. Taylor & Francis, London, pp 277–291

Rogers EM (2003) Diffusion of innovations, 5th edn. Free Press, New York

References

Sheth JN, Newman BI, Gross BL (1991) Why we buy what we buy: a theory of consumption values. J Bus Res 22(2):159–170

Sommerville I (2007) Software engineering, 8th edn. Pearson Education, Harlow

Spradley JP (1980) Participant observation. Holt, Rinehart and Winston, New York

Sward (2009) Leicester-partyrender-200903.png

Sweeney JC, Soutar GN (2001) Consumer perceived value: the development of a multiple item scale. J Retail 77(2):203–220. Available at: http://www.sciencedirect.com/science?_ob=MImg&_imagekey=B6W5D-435CJ1X-3-1&_cdi=6568&_user=122878&_orig=search&_coverDate=04/01/2001&_sk=999229997&view=c&wchp=dGLzVtb-zSkWA&md5=ae26ecd8678ca80010044805d8239184&ie=/sdarticle.pdf

University of Heidelberg (2010) Openrouteservice. OpenRouteService.org, 2010 (March 25th). Available at: http://openrouteservice.org/. Accessed 2 April 2013

Wood H (2009) MK Cake 1b.png. Available at: http://wiki.openstreetmap.org/wiki/File:MK_Cake_1b.png

Zeithaml VA (1988) Consumer perceptions of price, quality, and value: a means-end model and synthesis of evidence. J Mark 52(3):2–22

Chapter 4
Study Two: Understanding Design with VGI Using an Information Relevance Framework

4.1 Introduction

The inclusion of information by *potentially untrained volunteers* (VGI: Goodchild 2007) alongside that of the trained professional (Professional Geographic Information, PGI) has been one of the most significant shifts in the way information delivers meaning about our environment since the birth of Web 2.0 and neogeography. Whilst in their most basic forms VGI and PGI may be similar, it is the different ways in which these forms of information describe the environment—e.g. the structure of data and terminology used—where their variances are most prominent.

Individuals typically *search* for and *use* information, making choices whether to accept or reject discovered sources and deriving value from information based on its relevance to their needs (Tóth and Tomas 2011). In the context of *data quality* (Coote and Rackham 2008) and *User Centred Design* (Preece et al. 2002), design of new information delivery systems should be based on the users' *capabilities, current tasks and goals, conditions of product use* and *constraints on the product's performance*. Elwood (2008), alongside Zielstra and Zipf (2010) proposed that both VGI and PGI possess specific advantages and disadvantages for the *end user*, suggesting that no single information type may fulfil all of a user's requirements. Consequently, the development of mashups that utilise the best aspects of VGI and PGI have great potential to enrich the user experience when delivering information. Importantly, the work of these authors relates to the different levels of actual utility provided by data rather than the perceived utility derived from the resultant knowledge.

Within this book, the scoping study demonstrated that the perception of VGI is dependent on the particular use group, and the nature of their information use. To date the majority of research into the use of VGI has focused on the delivery of information through mobile, Global Positioning System (GPS) enabled devices, (Sun and Song 2009), the level of user trust in VGI by comparing it to PGI sources (Bishr and Janowicz 2010; Haklay et al. 2010) and objective *quality* within VGI

(Mummidi and Krumm 2008). This however does not address the differences in user perception of VGI and PGI, describing how one source is selected while another may be rejected. This is the topic this chapter aims to investigate.

4.2 Aims

The aim of this study was to take a user centred approach to studying the role that VGI plays when used alongside PGI within a realistic context. This included the utilisation of *information relevance* (outlined below) as the guiding theory for investigating how VGI and PGI is *perceived* and *used* by the study participants. The scientific rationale for this approach was that it enabled analysis of how information is actually used, and its potential application to a wider set of usage contexts. This was based on identifying key characteristics of the users and their tasks, and attributes of the information used.

It was the intention of this chapter to produce a greater understanding of effective use of VGI alongside PGI in the design of consumer orientated applications products and services. Therefore, the objectives of this study were to explore:

1. How VGI and PGI offer different benefits to the end user in a realistic scenario;
2. The strengths and weaknesses of VGI and PGI relative to how they meet the information requirements of the user's tasks and activities;
3. How VGI and PGI may be effectively integrated to produce highly usable and effective applications.

4.3 Study Rationale

4.3.1 Selection of Study Community

In order to investigate the perception of VGI and PGI in use, a user group was required that already made critical use of both VGI and PGI. The broad category of *Outdoor Adventure Recreation* was selected for the focus of this study due to the key role of geographic information (GI) within these activities. Importantly, outdoor adventure activities exhibit a relatively high potential for personal risk due to uncertainty and temporal variation in the conditions of the environment in which they participate (Ewert and Hollenhorst 1989). It was assumed that this relatively high level of uncertainty relating to environmental conditions (and the potentially serious consequences) would shape the accessing and use of information, and would encourage the participants to critically use a wide variety of information sources while being open to innovations where beneficial to them (Richins and Bloch 1986).

Communities were discovered through the 2009–2010 GeoVation Challenge (Ordnance Survey 2010), presenting business concepts for novel and use of GI.

4.3 Study Rationale

The relevance of such an approach was how those communities had a demonstrated and prominent need for information, not yet covered by traditional PGI. Therefore, the most suitable and prominent communities within this pool would have the greatest benefit to demonstrate the unique attributes and benefits of VGI in use.

Kayakers were selected as the participant community for this study due to their existing reliance on GI, use of dynamic information (e.g. river levels), dependence on multiple and varied information sources (e.g. books, blogs, etc.), range of potential experience levels and the potential of VGI to have influence on activities alongside PGI. Additionally while their sporting skills are specialist, their use of GI is an extension of those skills employed within normal/non-professional information searches. Therefore, the outcome of this research is scalable to the larger issues of how VGI may add benefit over and above PGI in other use contexts.

It is important to highlight here the relative complexity of the kayaking activity. As a sport, kayakers engage in training, small and large-scale river trips and social events. Within each of these activities, information in the form of internal and external information plays a crucial role in guiding the events in a safe manner. Therefore, it is essential that the tasks associated with these activities are understood, not for academic gain in describing the sport, but so information use (and the benefits of VGI and PGI) may be given their full and correct context of use.

4.3.2 Investigation Overview

This chapter comprises a multi-methods investigation into the support that VGI and PGI may provide for end users undertaking a specific task. Two independent investigations comprised (1) participatory observation to understand the social factors and interactions between users and (2) focus groups to gain a deep insight into the way groups of users utilise VGI and PGI. The qualitative research methods centred on understanding why different forms of information were used, how they were utilised and the way in which the characteristics of that information shaped the community's activities. Data was analysed through thematic analysis, with *relevance* used a theoretic framework. Results were analysed separately, but brought together in the discussion and conclusion.

4.4 Study Two A: Participatory Observation

4.4.1 Methods

4.4.1.1 Participant Sampling

To ensure a diverse representation of opinions a range of kayak clubs were involved in the focus groups, all adhering to the following criteria:

- Regular meetings between members in a formal location such as club or boathouse,
- Membership is open to the public, rather than being a private club,
- The main activities of the club are recreational kayaking, as opposed to slalom or racing,
- Regular trips are organised by the club members for other club members,
- A wide range of abilities included in the club, from beginner to expert.

4.4.1.2 Data Collection

During data collection, the position of *participant as observer was* sought (Gold 1969; Junker 1960). This was selected since it offered a useful degree of separation from the participants, not afforded by the more involved *complete participant*, yet enough involvement to gain a deep understanding of the issues difficult to obtain through the *marginal participant* perspective (Gold 1969). Participation took the following forms:

- Kayaking with club members on their weekly meetings
- Joining and training with the Loughborough Students Canoe Club (LSCC) throughout the study investigation period
- Kayaking river trips with clubs involved with this study.

Data for participatory observation was captured using *descriptive* observation under the dimensions highlighted in Table 4.1 to provide a rich and useful insight into user perceptions.

4.4.1.3 Data Analysis

McCall and Simmons (1969, p. 3) stated that the output from participatory observation is *an analytic description of a complex social organisation.* Records of observation were not coded, yet the statements and outcomes helped to validate and put into context the data from the focus groups.

Table 4.1 Dimensions of descriptive observation (Spradley 1980)

Descriptor	Definition
Space	Layout of the physical setting; room, outdoor spaces, etc.
Actors	The names and relevant details of the people involved
Activities	The various activities of the actors
Objects	Physical elements: furniture, etc.
Acts	Specific individual actions
Events	Particular occasions, e.g. meetings
Time	The sequence of events
Goals	What actors are attempting to accomplish
Feelings	Emotions in particular contexts

4.4.2 Results and Analysis

Participatory observation occurred on 12 occasions, with over 100 members from independent kayaking clubs; see Fig. 4.1.

The following key outcomes were derived from the observation during the study:

- Information serves to inform ideas about situations, critically analysed by participant based on past experience.
- Information is no substitute for experience; less experienced kayakers will seek to discuss issues with more experienced kayakers during an information search, and will value the opinions of their more experienced peers over third party information.
- The main role of information to the kayakers was allowing for the effective management of risk. Here, information was gathered up to the point where the participants felt they can kayak within the given risk conditions, creating a feeling a security.
- Activities centred on the social aspects of the sport, in some cases being seen as more important and prominent than the physical act of kayaking.

Fig. 4.1 Examples of participatory observation. **a** Very low water levels not predicted by VGI or PGI, **b** Unpredictable events, a split in a Kayak while on river, no emergency plan, **c–f** Engaging with participants during observation. Image **a** first published in Parker et al. (2012a), Image **b** and **f** first published Parker et al. (2012b)

During the observation (on the water) sessions, there was no evidence of participants consulting reference material or official guides. This suggests that external sources of PGI and VGI were used *during* the planning phase only. This was surprising, since it was assumed that guide books and similar would be used while kayaking. However, it was clear that environmental information cues, such as river levels and potential obstructions were actively sought, the main objective being the effective management (as opposed to *minimising*) of risk. These environmental cues clearly satisfy several of the relevance criteria *including accuracy, currency*, and *tangibility*. In addition, verification was also important, where multiple cues (e.g. relating to presence of obstructions) were sought. The role of experience of fellow kayakers was also key, in the search for (and interpretation of) external environmental cues.

4.5 Study Two B: Focus Groups

4.5.1 Methods

4.5.1.1 Participant Sampling

Non-probability purposive sampling methods were used to identify participants from the diverse range of kayaking clubs selected originally for participatory observation. The specific criteria for participant selection were:

- A minimum of 2 years kayaking experience
- Familiarity in planning of kayaking trips
- Experience using professional and amateur volunteer information sources
- Are not excluded from participation under ethical terms.

Participants in the focus groups were categorised by their number of years' experience kayaking as it was assumed that the *more experienced* kayakers may respond to information differently than *less experienced* kayakers. For analysis, kayakers of over 5 years' experience are referred to as *experienced*, whereas 1–4 years' experience counted as *intermediate*. Thirty-two participants (23 highly experienced, nine intermediate) from separate kayaking clubs took part in the four focus groups, and 50+ club members were involved anonymously in participatory observation. Although clubs had their own distinct focus (racing, white water, sea, social, flat water), all four were fundamentally recreational clubs. For their time and involvement in the focus group, the participants were offered an incentive of £5 per person, donated to the club.

4.5.1.2 Data Collection

In order to ensure the appropriate nature of the questions put to the focus groups, and the correct interpretation of their answers, focus groups were conducted after participatory observation.

As commented by Morgan (1998), exploratory studies require a less structured approach to the group interview than formal interviews where a *known* entity is being tested. Questions were used to guide the group discussion, yet allow enough leeway to develop the content of the discussion. In order to keep a scientific rigour a set question sheet was developed to offer the same basic questions to all focus groups. A series of thematic questions were devised in order to extract the desired information from the participants through engaging conversation and exploration of topics amongst participants (Krueger 1998b). Consequently, the questions centred on understanding:

- The information search process involved in planning of kayaking trips
- The positive and negative kayaking experiences of kayaking trips in relation to the impact of information
- The benefits of both amateur and professional information sources
- The nature of trust in information.

Sessions were recorded for later transcription, with group members being provided with additional material to make notes, sketches (etc.). The length of the session was not predefined, but tended to last for an hour.

4.5.1.3 Data Analysis

Thematic analysis was selected due to its focus on identifying themes and patterns in participant behaviour, and the development of deep insights in phenomena from which hypothesis and/or theory may be generated (Boyatzis 1998; Stake 1995; Yin 1994). From the work of Aronson (1994) and Boyatzis (1998), the following thematic analysis practice relating to this study was recognised:

1. *Developing Themes and Codes*—Combine and catalogue related patterns into sub-themes, producing a comprehensive story of their collective experience.
2. *Sensing Themes*—Patterns of Experience are collected from the data, recognising a code-able moment.
3. *Consistent and Reliable Coding*—Identify all data that relate to already classified patterns.
4. *Review of Codes*—allow for the coding structure to change with themes emerging from the data.
5. *Testing Reliability and Interpreting The Information*—Build a valid argument for choosing themes and formulate 'theme statements' to develop a 'story line'. When the literature is interwoven with the findings, the story that the interviewer constructs is one that stands with merit.

Figure 4.2 demonstrated the relation between the study objectives and theory used to guide the research and analysis within this chapter. From this, the basic categories used in coding were generated (Table 4.2).

4.5.1.4 Results and Analysis

Thirty-two participants took part in four focus group sessions. During the focus groups, PGI sources mentioned included guidebooks, maps and official reports, with VGI focusing mainly on forums, amateur reports and social media. A detailed coding of the focus group—including the number of references made to each theme and the number of participants who mentioned that theme—enabled an investigation of the relative importance of the information relevance attributes and a comparison between VGI and PGI.

Krueger (1998a) highlights *frequency, extensiveness* and *intensity* of participant comments as the key to understanding their general importance. For this reason the results within this section presents both the frequency of coding references, and the number of participants who voiced opinion on that subject. The intensity to which phenomenon was expressed during the focus group is considered during the analysis phase.

4.5.1.5 Hierarchical Task Analysis

In order to contextualise the impact of VGI and PGI across the entire trip process, a *Hierarchical Task Analysis* (HTA) was performed based on the focus groups and participatory observation. The HTA was required in order to identify which activities are more likely to draw on external information sources and to provide a framework for understanding the roles and influence of VGI and PGI.

Figure 4.2 was drawn from analysis of all data collection methods used. The HTA was developed to demonstrate the decomposition of goals, their relation to information types and the information required to execute each stage. Of this, the two categories of information considered (as describing an impact on the user in terms of their information needs) were *declarative* and *procedural*. Here, declarative information relates to information which must be understood and retained, whereas procedural information is the delivery of instruction (Ummelen 1997).

After the first draft of the HTA was developed, reliability was assessed through two additional focus groups involving experienced kayakers at Rugby and Rutland Water Canoe Clubs. Participants were sourced through the same methodology as in the main focus groups. Following discussion of the draft HTA, amendments to the structure, process and description were made as required.

Further description to Fig. 4.2 and the four levels of the HTA are described in Table 4.3.

Analysis of the focus group transcripts with reference to information use demonstrated that personal experience is used as a filter for volunteered and

4.5 Study Two B: Focus Groups

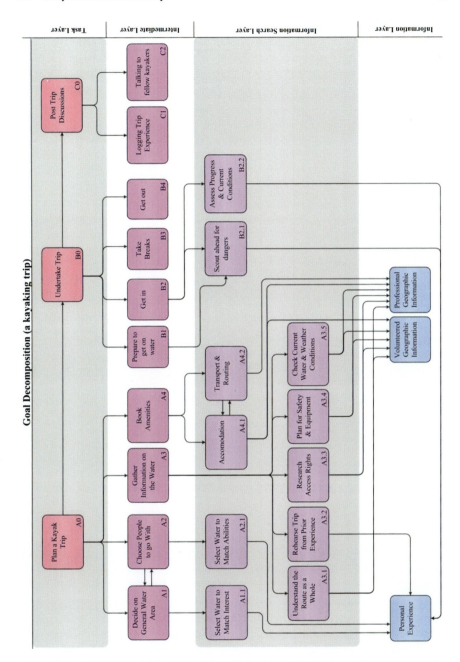

Fig. 4.2 A hierarchical task analysis of a kayaking trip. based on Bhavnani and Bates (2002), first published in Parker et al. (2012a)

Table 4.2 Outline of coding scheme used within the study

Study objectives	Guiding theory	Coding category	Sub-category
1. *How VGI and PGI offer different benefits to the end user in a real world scenario*	Information can benefit users in one of three stages of activity: planning, doing and reflecting (Davis 2005; Gitelson and Crompton 1983; Money and Crotts 2003)	Impact on trip activities	Planning Undertaking trip Post discussion
2. *The strengths and weaknesses of VGI and PGI relative to how they meet the information requirements of the users' tasks and activities*	Dissemination of information sources on unknown destinations (Gitelson and Crompton 1983; Hawkins et al. 1995; Weiss and Heide 1993)	Source of information	Formal Informal
	Professionalism is relative to authority of source (Coleman et al. 2009)	Identification of volunteered and professional information	Volunteered Professional
3. *How VGI and PGI may be effectively integrated to produce highly usable and effective applications*	*Relevance* of information to the user (Alonso et al. 2008; Barry and Schamber 1998; Cooper 1971)	Information	characteristics Accessibility Accuracy Affectiveness Availability Clarity Currency Depth Quality Tangibility
Verification			

professional information. This observation is mirrored in the *declarative information* column. Here, rather than requiring information in order to execute the various goals of the planning process, the participants require a certain degree of personal experience in order to fully complete the planning process.

An interesting outcome from the HTA generation was task C0—Post Trip Discussions. When asked about the trip experience or activities, no responses were made towards activities after their time on the water. However, participants placed high relevance on interpersonal communication, using their friends and social networks as efficient and effective data sources. While this does not constitute VGI due to its very limited potential to be shared with a large audience, it demonstrates the desire to share information and experiences which is at the heart of VGI creation (Feick and Roche 2010; Goodchild 2008; Scharl and Tochtermann 2007). However, it also demonstrates how the participants did not

4.5 Study Two B: Focus Groups

Table 4.3 Declarative and procedural relations to the HTA information layers

Layer	Declarative	Procedural
Task		*Task selection* First plan the trip, and then embark on it
Intermediate	*Existence of* 1. Experience kayaking 2. Judging information sources 3. Making bookings 4. Judging river conditions	*Strategy selection* Conduct each strategy in sequence to build up a knowledge base used for conducting a kayak trip *Method for planning a trip*: 1. Decide on general river area 2. Choose people to go with 3. Gather information on river 4. Book amenities
Information search	*Existence of* 1. Judging water conditions 2. Considering experience of others 3. Considering multiple information sources and converge on 'truth' 4. Organisation skills 5. Information search skills 6. Communication skills	*Command selection* 1. All information search options should be completed sequentially as indicated by their numerical indicator 2. If a stretch of river has been predetermined, then only search options A3.1–A3.5 should be completed 3. Any of the search options provide information which would endanger trip members, return to A1 OR cancel trip 4. Options A4.1 and A4.2 continue in iteration until amenities and loGIStics are organised
Information	*Sources of VGI*: 1. User generated river guides 2. Kayaking websites 3. Local river guides 4. Word of mouth 5. Social media *Sources of Professional Information*: 1. Kayaking guidebooks 2. Maps 3. Official data websites 4. Tourist information 5. River access officers	*Information types to utilise*: 1. Personal experience 2. VGI 3. Professional information Use is relative to the *information search* activity being engaged

see this activity as highly important or relevant, limiting the potential of this information.

Stage 1: Planning

As shown in the HTA, in the earliest stages of the kayaking activity (A1–A2), *internal information* in the form of personal experience is the predominant information source. This was supported by the participatory observations made during the trip. For example, Fig. 4.1 shows the water levels at the get in point, a water measure and a prominent bridge. While such measures may be categorised as *official*, it was the participant's internal knowledge and experience that gave those features meaning rather than information acquired prior to the trip. Table 4.4 gives an overview of the outcomes of the focus groups, related quotes from the participants to support an overview of information use during trip planning.

Stage 2: Undertaking

The majority of responses made in reference to the impact on information on kayaking activities were in the context of the trip itself. Table 4.5 gives an overview of the outcomes of the focus groups, related quotes from the participants to support an overview of information use during the kayak trip.

Stage 3: Post Trip Discussion

Table 4.6

4.5.1.6 Relevance of Information Sources

Accessibility Table 4.7
  Accuracy Table 4.8
  Affectiveness Table 4.9
  Availability Table 4.10
  Clarity Table 4.11
  Currency Table 4.12
  Depth Table 4.13
  Quality Table 4.14
  Tangibility Table 4.15
  Verification Table 4.16

4.5.1.7 Sources of External Information

Information identified as PGI was more likely to be perceived as out of date, while VGI had a higher tendency to reflect current conditions. However, this is not a reflection of the level of professionalism (or amateurism), but is due to the typical channels of delivery of these types of information. PGI predominantly comes from formal sources such as printed media, while volunteered information comes from informal (and particularly online and face-to-face) sources. The most prominent informal (VGI) sources are expressions of people's personal experience through either word of mouth, or online discussion groups. Additionally, the features used to assess the conditions of the water during kayak trips may be official landmarks such as a water gauge, but require personal experience to understand and make use of these information cues.

4.5 Study Two B: Focus Groups

Table 4.4 Outcomes relating to planning

Outcome	Quote	Comment
The only theme in the results relating to information use and its accessibility was the negative response that professional information is hard to obtain (four cases, four references)	I found trying to get hold of access information can be quite hard. Quite often, I just resign myself to just asking a mate who's been there recently, or perhaps someone who is more in the know than me [#2-1-07]	The general low support for this comment suggests participants did not have trouble accessing the quantity of information they required for trip planning
Some comment was made on the usefulness VGI and PGI. However, this was not a highly supported comment, and was voiced only by experienced kayakers for whom part of the excitement of kayaking is adventure and discovery	I always read the guidebook just before, someone's driving me to the river you have a quick flick through it, and then when you are put on the water you only forget what was in there anyway; you just go with it [#2-2-04]	
PGI provides details about the general topography of the river (ten cases, 13 references), while VGI provides information on specific points of interest (11 cases, 16 references)		The issue of resolution of information[a] may play an important role in its *relevance* to the user
A relatively small sample (six cases, ten references) on how VGI needs to be checked before use	VGI must be verified somehow [#2-4-05]	Made at a much higher rate than the need for participants to check PGI (one case, one reference). This is of particular interest when considering the high number of references made for VGI being up to date, and PGI out of date

(continued)

Table 4.4 (continued)

Outcome	Quote	Comment
Personal experience allows critical evaluation of external information sources (4 cases, 6 references)	Often rafters are also kayakers… but again you still have to take it with a pinch of salt… so what's easy for them isn't necessarily easy for you. So you still got to read the guidebook, scout as much of the river as you can as possible [#2-1-07]	Highlights an important factor that external information searches alone cannot provide for a successful user experience (a successful kayak trip)
Very high responses were given to both VGI (21 cases, 42 references) and PGI (23 cases, 51 references) sources enhancing the participant's general understanding		Suggests that the most relevant factor of information in the activity is its ability to enhance the understanding of the user

[a] *Resolution*—wide geographic range with low individual item detail equates to low resolution (but high coverage in m^2), low geographic range with high individual item detail equates to high resolution (but low coverage in m^2)

4.5 Study Two B: Focus Groups

Table 4.5 Outcomes relating to undertaking

Outcome	Quote	Comment
The highest response was for personal experience being a useful tool in helping the participant to engage in further personal experiences. A significant amount of participants (17 cases, 28 references) commented that it is the adrenaline or challenge of the sport that they find highly enjoyable	No one else has really seen it other than the people that have been down the river. And if you get to, you can get certain places that you'd never be able to walk to [#2-1-08]	This collection of positive experiences are enabled by personal experience, and while information may have enabled the trip to happen no information provides the positive experiences which the participants enjoy
While information does not provide the participant with positive experiences, the negative aspects of the tangible outcomes demonstrate that a lack of information may allow for negative experiences	We were in Austria and we were driving along a road. 'Oh that looks like a good rapid'… didn't scout it… It was just ridiculously steep and just huge holes. it was a blur [#2-4-03]	This section suggests that information does not provide the kayaking participants with good experiences, but it can prevent them from having bad experiences and thus enables an enjoyable trip to occur
Only five cases with six references made note that professional information sometimes provided incorrect information, against one case and one reference for VGI	We rang up the river information office and I said "what's the levels like?" and he said "very favourable". And when we got there we had to walk around half of it was so low! I was like 'if this is favourable….' [#2-2-01]	Related to the high proportion of participants who commented that professional information has a tendency to be out of date. This suggests that the more up to date the information is the more likely it is to reflect the current conditions and thus be correct

Table 4.6 Outcomes relating to post trip discussion

Outcome	Quote	Comment
The salience of VGI during the trip planning stage suggests dissemination and volunteering of information post trip to other kayakers *is* a key element of the trip activities	By chatting to paddling friends, I usually can decide whether the particular river is within my comfort zone and abilities [#2-2-04]	
This dissemination process may be formal processes but mostly they are informal *chatting* to other kayakers in informal settings	If you're padding a stretch of river there's generally certain points you can get on… and there's always a pub along that stretch at some point. So if you see other paddlers you talk to them [#2-1-02]	

Table 4.7 Relevance of information sources: accessibility

Outcome	Quote	Comment
Although one of the key characteristics of PGI is its premium association[a], only five participants mentioned this as a problem accessing information, concentrating on the cost of PGI being an inconvenience rather than a preventing factor	Quality of info varies a lot and you need to pay before you see what you get [#2-3-09]	Paying for information is not seen as a great burden on the information seeker, more a way of passage to the information that will make their trip success; if proprietary information is sought
Very high salience (29 cases with 77 references) was given to VGI from kayaking websites and forums. In addition to this, only four participants made five references towards VGI being free	People don't generally want money for it [#2-3-09]	This suggests that while free information is of benefit to the user, it is not a factor which makes the information appear more attractive to the user

[a] The professional(s) *selling* their *certified* information as a source of income (O'Brien 2010)

Table 4.8 Relevance of information sources: accuracy

Outcome	Quote	Comment
While more references were made to VGI sources being accurate than were made towards PGI sources, more VGI sources are used in the convergence of truth than PGI sources	Multiple sources converge on truth rather than hold truth within a single source [#2-1-04]	Accuracy of the information being received is an important factor, but it is most important when considering multiple sources and factors which can confirm or reject the statements made
Through the research no comment was made on the emotional connection between the participant and professional information. Instead, the only emotional connection was down to persons encountered during trips which falls outside the remit of VGI	They always seem to be having a worse day than us though......the fourth one [fisherman] jumped up and down, looked miserable, looked like we have ruined his whole day, we just laughed [#2-2-03]	This suggests that while affection towards an information source may influence the sources a user goes to in their information search, it does not influence their general preference for use of PGI or VGI

Table 4.9 Relevance of information sources: affectiveness

Outcome	Comment
Very high salience can be assigned to the use of VGI from kayaking websites and forums, yet limited comment was made by participants about the volume of information available	Either the participants are not overly concerned with the volume of information available, or the information sources available fulfil their need. One explanation may be some kayakers enjoy the sense of the unknown, and therefore a lack of information may add to the user experience

Table 4.10 Relevance of information sources: availability

Outcome	Quote	Comment
A salient number of participants (nine cases with 12 references) commented that they found PGI at time vague and hard to understand	If you're reading it out of a book you might not quite understand certain aspects [#2-3-06]	VGI offers a certain degree of clarity above that of professional information. This may be because most if not all of the VGI relevant to kayakers comes from homogenous sources, and thus should be easier for the information seeker to ingest

Table 4.11 Relevance of information sources: clarity

Outcome	Quote	Comment
Five cased with eight references noted that professional information was in general well structured	[It's] often produced in a more usable format and more accessible (published bodies/websites), not trawling through information on forums [#2-4-05]	While these outcomes may suggest that PGI has a communicatory advantage over VGI in terms of clarity, the lack of comment towards VGI makes it difficult to state a definite outcome in terms of relative strengths and weaknesses
A salient portion (17 cases with 30 references) commented that professional information tends to be out of date	What maps and guidebooks don't give you is up to date information. Just because it was a good guide to the river five years ago doesn't mean it's a good guide to the river now [#2-1-05]	

4.6 Discussion

4.6.1 Impact of Information Depth and Scope in Understanding the Outdoor Environment

Analysis of VGI and PGI according to the relevance framework of Barry and Schamber (1998) has shown some clear differences in the perception of these information sources by end-users. This study demonstrated that PGI has a lower degree of perceived overall depth about specific locations than VGI, but a greater degree of overall scope and consistency of coverage. The participatory observations showed that when the users talked about VGI sources, the topics covered were also of greater diversity than their PGI counterparts. Consequently, PGI provides information on the general, wide reaching topography, while VGI

Table 4.12 Relevance of information sources: currency

Outcome	Quote	Comment
Three cases with four references commented that VGI was out of date. This is enhanced by the highly salient proportion of participants (16 cases with 23 references) who felt that VGI was up to date	You get things like 'trees' across big rivers' and things like that. Quite often within a few days you will get a notice on a forum saying "be careful there is a big tree stuck on the rock on 'this' bend" sort of thing [#2-3-09]	However, only six percent of participants referred to professional information being up to date
Information needs to reflects the conditions of the outdoor environment when the participant experiences it. The importance of this is highlighted by the information sources which can capture rapidly changing and largely unpredictable factors (such as river conditions) being seen as more accurate than slower responding sources;	[VGI is] often more accurate with [the inclusion of] real time information [#2-1-02]	Inferring the paddleable conditions of a river and reporting them through VGI channels as observed during participation demonstrated VGI's unique ability in delivering this need compared with the planned surveying practices of PGI
Although a proportion of participants made comment that VGI is incomplete (four cases with eight references), a far greater salience can be given to participants perceiving professional information as incomplete (11 cases with 22 references)	Like we said with maps, you can't gauge, like I said, bank levels, and you can't, it's, there more for distances and everything like that [#2-1-08]	The focus groups also suggested that PGI can (at times) describe the general overview of the outdoor environment, yet misses key details about the features most important to the participants

provides detail about specific locations, sometimes in much greater detail, but with patchy coverage. These elements may be considered as intrinsic to the scope and level of detail in the GI, which as Levitin and Redman (1995) suggested, are important dimensions of data *quality*. This (alongside *price* and *value*) is one of the key criteria for product selection (Zeithaml 1988). Quality judgements in relation to information-based products are therefore important in terms of their adoption by potential consumers.

Unexpectedly, low-level dissatisfaction with PGI due to incompleteness relative to the needs of the users was evident within the focus groups. Consequently, a need exists to understand the user's information needs further, and then tailor the information provided to fit these needs. Ivergård (1982) commented that users' reactions to information are typically in relation to the amount of information *expected* rather than the amount *actually* found. Ivergård's comment may explain

4.6 Discussion

Table 4.13 Relevance of information sources: depth

Outcome	Quote	Comment
Rather than utilise a single VGI source, they access multiple sources and converge on the truth	I think you use it, all these little bits of information to build a whole picture of what you want to do [#2-1-04]	This contrasts with use of PGI—with only 9 % of participants stating that they would use multiple sources of information rather than use a single PGI source. Regular emphasis was used by participants to stress the importance of using information to confirm discovered VGI
An almost equally strong resonance (11 cases with 26 references for amateur volunteered, 13 cases with 22 references for professional) was perceived by the participants that the information they receive is opinionated or subjective depending on the originator	[VGI] It's very open to interpretation. Someone else's grade 5 can be someone else's grade 3 [#2-4-05]	Kayakers use predominantly personal experience while on the water. Personal experience acts as a filter for information use while planning a trip. This suggests that the information seeker is subjective, in that what they consider to be difficult is personal to them and thus they must understand the conditions of the water being described to match it to their understanding of difficulty rather than take the *as stated* level of difficulty. This may explain why both VGI and PGI are seen as subjective in the eyes of the information seeker

the level of dissatisfaction with PGI sources, which (as this study has shown) are seen as having broad scope, but are perceived as incomplete in relation to contextual detail.

Personal experience is influential in how it enables analysis and validation of external information sources. This outcome is consistent with the work of Xiang and Gretzel (2010) who demonstrated how when planning tourism activities people already utilise *social media* to advise them on their activities once they have decided on the general location of their trip. Consideration should be given here to the information sources mentioned within this chapter. While focus groups reduce the degree to which important information may be overlooked, certain information types may lend themselves to being mentioned more frequently than others. For example, forums can be considered vast repositories of information, and thus worthy of mentioning. However, less established sources such as video

Table 4.14 Relevance of information sources: quality

Outcome	Quote	Comment
An interesting outcome from the data was a proportion of the participants (14 cases, 18 references) said VGI was unreliable	Locals will probably know more about access, but locals are often not kayakers [#2-1-07]	Although there was relatively limited reference to VGI being *purposefully misguided* or otherwise unreliable, its presence indicated a level of distrust in the potential quality of VGI. This suggests that for volunteered systems to be seen with confidence from a user case, a mechanism is required to overcome this perceived sacrifice in obtaining and using VGI
A number of participants (11 cases with 23 references) perceived VGI reliable	I think it's possibly more reliable, up to date, and you could be talking to somebody who is local and knows the river and walked past it that morning [#2-1-02]	What is also interesting but not unexpected is that a larger proportion of participants (19 cases with 33 references) perceived professional information as reliable. Participants in the focus groups commented that PGI creators are *"honest and trying to the best of their knowledge; it's their reputation"* [#2-4-03] and their material is *"usually [a] very trustworthy source with high level of experience"* [#2-4-01]
A proportion of the participants commented that they trust their personal contacts more than anonymous sources such as guide book authors or forum posters	I'll chat to my friend and he will say 'yeah your able to do that'... Whereas if he said 'ooh', I'm not doing it. I put that much sway on what he says that it really does influence where I want to go, what I want to do... you just done get from websites or books [#2-2-04]	This suggests that factors such as social networks, homogeneity and interpersonal trust in the information originator may be key factors in the information seekers perception of the information's quality

websites (e.g. www.youtube.com) may not be seen as important or formal enough, so not mentioned. Therefore, it is important to consider all outcomes within this study relating to utterance of sources as indicatory, rather than as a measure of importance or prevalence.

As highlighted by Manchala (2000) the user's overall experience of interacting with information is dependent on trust and the user's willingness to utilise the

4.6 Discussion

Table 4.15 Relevance of information sources: tangibility

Outcome	Quote	Comment
Additionally, the forms of information that may be authoritative (e.g. Ordnance Survey) may not be able to report changes in the environment at a fast enough rate to be considered tangible	OS Maps - out of data if in paper and costly [#2-2-06]	Demonstrates how the information kayakers rely in the most relates to the fast changing environment (e.g. water levels) rather than static features (e.g. hills). This is unique to kayakers
Although of low frequency (four cases with five references across all information types), comment was made that both professional and VGI are no substitute for experience	And at the end of the day you have to have faith in your own ability, either as a team or as a paddler as to what you're going to do or what you're not going to do, because with all the best information in the world you're not going to know until you get there [#2-1-04]	The intangible, personal experience plays a more prominent role in the kayaking activity than the tangible external information from VGI or PGI sources

Table 4.16 Relevance of information sources: verification

Outcome	Quote	Comment
Participants commented that personal experience and other people sharing their personal experience is the best form of information; above professional and anonymous VGI	I think people that have done the river before are the best people to talk to. They know your level of paddling ability and if they think 'oh no, it's not for you', they'll say 'it's a great river… but I don't think you're at that level yet' [#2-1-08]	These outcomes suggest that while third party information sources are vital to the planning process, they may not make up 100 % of the information search process. While this study does not conclude if these interpersonal communications are necessary in an information search context, they are of high importance and lead weight to the concept that *complete* mixed source information sets should contain volunteered, professional and interpersonal elements in order to produce a highly effective and satisfying solution to the end user

information in future instances. Consequently, if VGI is utilised alongside PGI in applications in such a way as to increase the positive experiences for the user then the *trust* perceived by the user towards the application may be increased.

This research has demonstrated how the different dimensions of user perception (e.g. accessibility, accuracy, etc.) relate to their overall trust in the information. This is a useful development in the overall understanding and application of VGI relative to the work of Mummidi and Krumm (2008) in the need for objective quality in VGI. Consequently, the *depth* and *scope* of the information sources are most important to the user when searching for trip planning information. Although the *completeness* of individual information sources is important, it is more important that the whole collection of information sources (i.e. VGI and PGI together) produce a *complete* image when they are combined and considered alongside each other. Additionally, this may be in line with Grira et al. (2010) who demonstrated that by including the contributions of amateur volunteers a GIS may improve its overall objective quality.

The level of precision in the explanation of the outdoor environment can be considered alongside the work of Corona and Winter (2001, p. 1) who commented that *people that move in unfamiliar environments need precise instructions to reach a specific location*. It may be expected that the more precise information the user requires, the higher the potential dissatisfaction with PGI may be felt. This presents a great opportunity for VGI to be a highly usable form of information to the user; being *effective, efficient* and *satisfying* (ISO 9241-11 1998). This however, may only held true if VGI can be demonstrated to provide the highly precise and detailed descriptions of specific points in the outdoor environment—as suggested by this study.

This discussion highlights that the *depth* and *scope* of the information sources are important to the user when searching for trip planning information. Although the completeness of *individual* information sources is important, it is more critical that the collection of information sources (e.g. all recent posts on all kayaking forums) produce a *complete* image when they are combined and that all are considered relative to the time frame of their origin.

4.6.2 Influence of Information Currency

This study highlighted how VGI sources were preferred in situations where the geographic features being described altered regularly (e.g. water levels). In contrast, PGI sources were preferred when describing relatively static geography (e.g. topography). It was clear from this study that the extensive use of VGI and its perceived usefulness is due to its currency; i.e. the ability for it to reflect recent changes within the application domain. These findings are in agreement with various authors (Nolan 1976; Gitelson and Crompton 1983; Schuett 1993) who demonstrated that in recreational environments information received from *informal sources* can be the most informative due to its ability to reflect changes in the environment.

This is not simply due to the volunteered nature of the information, but critically is also influenced by the channels through which VGI tends to be communicated. Information collected and distributed through regularly updated, interactive channels (rather than through the slower mediums such as print with longer refresh cycles) has a higher chance of reflecting current conditions, and satisfying the *currency* requirement within the relevance framework of Barry and Schamber (1998). The finding that VGI is best suited for fast changing *geography* that may be hard to capture through traditional methods is directly in line with the concepts outlined by Goodchild (2007) when he defined the term *Volunteered Geographic Information*.

An interesting consideration is the degree of *information redundancy* inherent in traditional PGI systems: the inclusion of non-essential information from the user's perspective (Badenoch et al. 1994). Since Ivergård (1982) commented that users react to information in relation to the amount of information *expected* rather than the amount actually found, an additional perspective on the user may be gained. In particular, when the user expects the information to reflect the current conditions, yet that expectation is not met, the abundance of non-essential additional information in PGI may have a negative impact on the user experience.

4.6.3 Importance of Real Time Information

One of the most unexpected findings from the study was the lack of either actual or desired access to GI in *real-time* while undertaking the kayaking trip. The kayaking environment itself presents challenges to information access: in particular the water-based environment and the lack of free hands. To date, much geographical user research has focussed on the delivery of location-based information; e.g., delivery to mobile phones (Sun and Song 2009; Tsou and Yanow 2010; Xiaolong 2007). However, the findings from this study question the extent to which such real-time information is useful, and instead suggest that when users are actually engaging with the environment, they are not necessarily motivated to find out more about geographical features but instead draw on internal information derived from their personal experience or direct communication of relevant facts from fellow participants.

This is shown in the records of participatory observation, where the members of the kayaking trip would look to the leader for guidance and advice, who in turn would rely on his personal experience and internal knowledge. Additionally, as highlighted by Arnould and Price (1993) this observation may be explained by the kayakers' desire for *river magic*, or a hedonic experience coming from the adventure of overcoming risk rather than simply engaging in kayaking on a river. Further generalizability of this may be seen in fields such as general tourism (Gursoy and Chen 2000) and store shopping (Cox et al. 2007), where the lack of complete knowledge (creating a degree of uncertainty) provides opportunities for uncertainty, and thus discovery leading to enjoyment. This outcome highlights how VGI has the greatest potential to impact on the outcome of the information-seeking user during the planning (rather than the activity) phase.

4.6.4 Importance of Information Access

Although VGI is often distributed under a *Creative Commons* licence—and is therefore free to access (Goodchild 2008)—this does not make it appear more appealing to the user; or to make PGI comparatively less attractive. The focus groups showed that participants used whichever information source they felt most likely to solve their information needs; be it either free as in a forum or at cost at in a book. This may be explained by the work of Richins and Bloch (1986) who asserted that the higher the perceived risk, the higher the involvement in the information search. This would suggest that individuals are more willing to spend resources (effort and/or money) for information if there is risk associated with an activity. As Borlund (2003) commented that the relevance of a document should be judged on the basis of its content rather than its physical properties, such as physical availability or monetary cost, which would explain this use of PGI. The finding that participants would pay for information if it was seen as appropriate and useful is interesting, partly due to the fact that proponents of VGI hold the free nature of their information up as a key reason why VGI is *better* and more *appropriate* for general use than PGI (Flanagin and Metzger 2008). Consequently, there may be an inverse relationship between an activity's risk and importance of the accessibility attribute -including the free nature of VGI.

4.6.5 Importance of Trust in Information

Participants used multiple sources of information to converge on truth rather than take single information sets as true. However, the multiplicity of sources used is not a direct indicator of their importance or impact, so further insight into the user judgements is required. Additionally, this section may be seen within the context of selecting information to fulfil a given purpose of the user. As described by Wang and Soergel (1998) in the context of document selection this is the final stage of user judgement in deciding if an information item should be used or not, following processing of information elements and combining of criteria.

As demonstrated within Sect. 4.5.1.6 (*Relevance of Information Sources*), personal contacts are a more trusted group than any other information source. This is a mirror of the work by Manning and Lime (1999) that many sources of information are used by outdoor recreation visitors for trip planning. Additionally, they demonstrated that these sources were not directly produced by management

4.6 Discussion

agencies (e.g. outdoor clubs, professional outfitters, guidebooks, newspaper, etc.) but by volunteers presenting their past experiences. This finding is also in line with observations by Rieh (2002) who pointed out that traditionally information search has focused on how accurately the topic the user is searching for matches the topic of the documents found, yet with online information searches people use diverse criteria of search topics simultaneously.

One explanation for the reliance on personal contacts more than professional information (as shown by this study) is offered by Schuett (1993), who suggested that the inherent risk involved in outdoor adventure activities may be the main reason for the use of more personalised sources such as *friends, outdoor stores*, and *professional outfitters*. Schuett also commented that friends and family are easier to get hold of, and because of interpersonal relationships already have an inherent measure of trust and reliability, which the consumer does not exhibit for the non-personal information sources. This is in line with Beatty and Smith (1987) who within the wider context of consumer product purchases noted that friends and family are consistently reliable sources for information.

However, this reliance on interpersonal relations in an information search environment is somewhat at odds with the comments of Rieh (2002) that web users' judgments of *quality* and *authority* are influenced more by institutional level of source (e.g. source reputation, type of source, and URL domain type) than by the individual level (e.g. author/creator credentials).

As shown in Sect. 4.5.1.6 (*Relevance of Information Sources*) the more *knowledgeable* and *accurate* an information source is (in the sense of reflecting the conditions of reality in line with how the information searcher will experience them), the more likely it is to be seen as authoritative and professional. In this situation, it is accuracy that might be inferring professionalism to the users, rather than a professional *label* emphasising accuracy. Importantly, professionalism in this context refers to the quality of the work rather than the credentials of the author. Additionally, accuracy can only be asserted after the information use event, and thus demonstrates the need for a feedback loop within the user/contributor context. If this was engaged with, it is possible that such a function may lead to increased judgements of professionalism in the data over time.

This may be explored further through the concept of *cognitive authority*, defined by Wilson (1983) as influences that a user would recognize as proper because the information therein is thought to be credible and worthy of belief. The significance of this is highlighted by Rieh (2002)—that in contrast to information *quality* (the extent to which information is actually useful, good, current and accurate) *cognitive authority* is operationalised as to the extent to which users *think that they can trust* the information. Consequentially, for VGI use by kayakers, the *quality* of the information influences the *cognitive authority* exhibited by the information.

A further explanation for the participant's perception of cognitive authority was offered by Rieh (2002), who observed that when academic participants were presented with work that appeared academic, they perceived its cognitive authority to be higher than work that appeared less scholarly. It is however not clear whether this refers to scholarly as an indication of absolute quality, or as an indication of the homogeneity of the contributor and user of the info. This offers further opportunity for investigation into the link between VGI presentation within neogeography and its perceived authority.

The link between accuracy and cognitive authority may be explained by the work of Corona and Winter (2001), who commented that *people that move in unfamiliar environments need precise instructions to reach a specific location.* Within such unfamiliar environments as Kayakers interact with, information accuracy may become more important than other factors such as cost or diversity of content. Additionally, Rieh (2002) mentioned that if there are a number of information resources related to their topical interests, then the consumer would want to find *useful* and *appropriate* information, and would be likely to base their actions on the concept of quality and authority. This also links (1) the outcome that multiple sources of information are used to converge on the truth to (2) the critical analysis of utility in the information and ultimately, the impact on cognitive authority.

4.6.6 Volunteer Reporting of Activity Experiences

Within the kayaking community, *feeding back* of experiences via informal channels is crucial to the information search activities when planning trips, yet is not explicitly stated as an important activity. Therefore, there exists a lack of perceived need to more formally feed such experiences back to others through VGI channels.

The low importance placed on actively disseminating experiences gained during the trip means that within the kayaker community a vast pool of potential VGI within individuals' personal experiences exists that is not freely available and easily accessible to others. This repository of information may therefore be considered *sticky* (Luthje et al. 2005), where the *cost* of accessing such information is effectively *the ability to ask a question to the individual who holds it.* Without being in contact with that person, or knowing that they may hold such information their experiences are consequently inaccessible.

4.7 Conclusions

Through investigation, this study has addressed the study aims in the following ways:

1. *How VGI and PGI Offer Different Benefits to the End User in a Realistic Scenario*

This study has shown that within the context of outdoor recreation, the commonly held assumptions that VGI is inferior to PGI, and that the most beneficial,

4.7 Conclusions

accurate and useful GI can only come from professional sources is no longer correct. In describing the outdoor environment for special recreation interest, PGI is more likely to describe the general geography and conditions of wide reaching features while VGI comes from a convergence of amateur sources describing specific regions of interest. Consequently, the end-user seeking information may discover relatively high levels of detail about specific locations from VGI, related to one another through the general description of the environment derived from a PGI source. One of limitations identified with VGI has been the relative difficulty in tapping into experiences of users due to the reluctance of seeing contribution as an important part of the trip process.

2. *The Strengths and Weaknesses of VGI and PGI Relative to How They Meet the Information Requirements of the User's Tasks and Activities*

The verification of VGI and the quality of the source are critical issues that influence the extent to which VGI is deemed relevant by a user. In discovering information about the outdoor environment that is not understood through internal information, verification can be achieved by reference to multiple sources that *converge on the truth*. Quality of source may come from knowing (and understanding the significance of) the credentials of the contributor.

From forums, websites and community notice boards, VGI was shown to be easy to access while offering a wide spatial coverage of potentially up-to-date information on geographic regions important to the disseminating community. Although it can be influenced by subjective interpretation from contributors it was generally considered reliable and relevant by participants.

It is more useful to consider the attributes of information (e.g. the update rate, ease of access) than just the level of professionalism of the author; i.e. whether it is VGI or PGI. This brings into question the practicality of the terms VGI and PGI in describing the usefulness of information from different sources.

The greatest opportunity for VGI to impact on outdoor activities is in situations where the current conditions of the geographic area are either not accessible via traditional cartographic means, are not sufficiently predictable through scientific methods, or are likely to have changed since they were last reported.

3. *How VGI and PGI May Be Effectively Integrated to Produce Highly Usable and Effective Applications*

The study suggests great potential for VGI to counteract the shortcomings of PGI sources in relation to the needs of the user. The integration of these two forms of data within a mashup could combine the structure, consistency and source quality of PGI with the currency and intuitive appeal of VGI. Such mashups would have higher personal relevance than could be achieved by either VGI or PGI alone.

This study has focused on kayaking, yet it points towards a significant opportunity for increasing the usability of GI by integrating volunteer and professional sources in other contexts. Developers of future GIS could maximise the synergy of VGI and PGI through understanding how different characteristics of each source can be used together to meet the needs of specific user groups and use contexts. The implication for those wishing to combine VGI and PGI when designing applications is to consider both information sets not as simply *volunteer* or *professional*, but as two different yet equally valid information sets within the rich tapestry of GIS.

In order to understand how the outcomes of this chapter may be applied to the wider range of consumers it is important that further research is undertaken with a different yet comparable consumer group to kayakers. Additionally, it is important that further research may add additional context to the outcomes of Study Two by focusing on the reactive perceptions of users to VGI during an information search.

References

Alonso O, Rose DE, Stewart B (2008) Crowdsourcing for relevance evaluation. ACM SIGIR Forum 42(2):9–15
Arnould EJ, Price LL (1993) River magic: extraordinary experience and the extended service encounter. J Consum Res 20(1):24–45
Aronson J (1994) A pragmatic view of thematic analysis. Qual Rep 2(1):1–3
Badenoch D et al (1994) The value of information. In: Feeney M, Grieves M (eds) The value and impact of information. Bowker-Saur Limited, Chippenham, pp 9–78
Barry CL, Schamber L (1998) Users' criteria for relevance evaluation: a cross-situational comparison. Inf Process Manage 34(2/3):219–236
Beatty SE, Smith SM (1987) External search effort: an investigation across several product categories. J Consum Res 1:83–95
Bhavnani SK, Bates MJ (2002) Separating the knowledge layers: cognitive analysis of search knowledge through hierarchical goal decompositions. Proc J Am Soc Inform Sci Technol 39(1):204–213
Bishr M, Janowicz K (2010) Can we trust information?—the case of volunteered geographic information. In: Devaraju A et al (eds) Towards digital earth: search, discover and share geospatial data. Workshop at Future Internet Symposium, Berlin
Borlund P (2003) The concept of relevance in IR. J Am Soc Inform Sci Technol 54(10):913–925
Boyatzis RE (1998) Transforming qualitative information: thematic analysis and code development. Sage Publications Inc, Thousand Oaks
Coleman DJ, Georgiadou Y, Labonte J (2009) Volunteered geographic information: the nature and motivation of produsers. Int J Spat Data Infrastruct Res 4:332–358
Cooper WS (1971) A definition of relevance for information retrieval. Inf Storage Retrieval 1:19–37
Coote A, Rackham L (2008) Neogeography data quality—is it an issue? In: Holcroft C (ed) Proceedings of AGI Geocommunity'08. Association for Geographic Information (AGI), Stratford-Upon-Avon, UK, p 1. Available at: http://www.agi.org.uk/SITE/UPLOAD/DOCUMENT/Events/AGI2008/Papers/AndyCoote.pdf
Corona B, Winter S (2001) Navigation information for pedestrians from city maps. In: GI in Europe: Integrative–Interoperable–Interactive, Proceedings of the 4th AGILE conference on geographic information science. Masaryk University Brno. Citeseer, pp 189–197
Cox AD, Cox D, Anderson RD (2007) Reassessing the pleasures of store shopping. J Bus Res 58(3):250–259
Davis C (2005) What's "going out" all about?. Loughborough University, Loughborough
Elwood S (2008) Volunteered geographic information: future research directions motivated by critical, participatory, and feminist GIS. GeoJournal 72:173–183

References

Ewert AW, Hollenhorst S (1989) Testing the adventure model: empirical support for a model of risk recreation participation. J Leisure Res 21(2):124–139

Feick R, Roche S (2010) Valuing volunteered geographic information (VGI): Opportunities and challenges arising from a new mode of GI use and production. In: Poplin A, Craglia M, Roche S (eds) GeoValue 2010 proceedings: 2nd workshop on value of geoinformation. Geovalue, HafenCity University Hamberg, Hamberg, DE, p 67

Flanagin AJ, Metzger MJ (2008) The credibility of volunteered geographic information. GeoJournal 72:137–148

Gitelson RJ, Crompton JL (1983) The planning horizons and sources of information used by pleasure vacationers. J Travel Res 21(3):2–7. Available at: http://jtr.sagepub.com/content/21/3/2.short

Gold RL (1969) Roles in sociological field observation. In: McCall GJ, Simmons JL (eds) Issues in participant observation: a text and reader. Addison-Wesley, Reading, pp 30–38

Goodchild MF (2007) Citizens as sensors: the world of volunteered geography. GeoJournal 69(4):211–221. Available at: http://www.springerlink.com/content/h013jk125081j628/

Goodchild MF (2008) Commentary: whither VGI? GeoJournal 72(3):239–244

Grira J, Bédard Y, Roche S (2010) Spatial data uncertainty in the VGI world: going from consumer to producer. Geomatica 64(1):61–72

Gursoy D, Chen JS (2000) Competitive analysis of cross cultural information search behavior. Tourism Manage 21(6):583–590

Haklay M, Ather A, Basiouka S (2010) How many volunteers does it take to map an area well? In: Haklay M, Morley J, Rahemtulla H (eds) Proceedings of the GIS research UK 18th annual conference. University College London, pp 193–196

Hawkins DI, Best RJ, Coney KA (1995) Consumer behavior: implications for marketing strategy 6th edn G. A. Churchill Jr., edn. McGraw-Hill Education, Chicago. Available at: http://books.google.co.uk/books?id=5G8LeHwN9-UC&q=Consumer+behavior:+Implications+for+marketing+strategy&dq=Consumer+behavior:+Implications+for+marketing+strategy&hl=en&sa=X&ei=hHQKUdKCGMec0AXwt4CACQ&ved=0CDEQ6AEwAA

ISO 9241-11 (1998) Ergonomic requirements for office work with visual display terminals (VDT)s—part 11 guidance on usability. International Standards Organisation, Geneva

Ivergård T (1982) Information ergonomics. Chartwell-Bratt Ltd, Sweden

Junker B (1960) Field work. University of Chicago Press, Chicago

Krueger RA (1998a) Developing questions for focus groups. Sage Publication Inc, Thousand Oaks

Krueger RA (1998b) Analyzing and reporting focus group results. Sage Publications, Thousand Oaks

Levitin A, Redman T (1995) Quality dimensions of a conceptual view. Inf Process Manage 31(1):81–88

Luthje C, Herstatt C, Von Hippel E (2005) User-innovators and local information: the case of mountain biking. Res Policy 34(6):951–965

Manchala DW (2000) E-commerce trust metrics and models. IEEE Internet Comput 4(2):36–44

Manning RE, Lime DW (1999) Defining and managing the quality of wilderness recreation experiences. In: McCool SF et al (eds) Wilderness science in a time of change conference (Volume 4: Wilderness Visitors, Experiences, and Visitor Management). USDA Forest Service, pp 13–52

McCall GJ, Simmons JL (1969) Issues in participant observation: a text and reader. Addison-Wesley, USA

Money RB, Crotts JC (2003) The effect of uncertainty avoidance on information search, planning, and purchases of international travel vacations. Tourism Manage 24(2):191–202

Morgan DL (1998) Planning focus groups. SAGE Publications inc, USA

Mummidi L, Krumm J (2008) Discovering points of interest from users' map annotations. GeoJournal 72:215–227

Nolan SDJ (1976) Tourists' use and evaluation of travel information sources: summary and conclusions. J Travel Res 14(3):6. Available at: http://jtr.sagepub.com/content/14/3/6.short

O'Brien OG (2010) London tube station usage. Homepage of oliver O'Brian, 2010 (Aug 31). Available at: http://oobrien.com/vis/tube/. Accessed 31 Oct 2010

Ordnance Survey (2010) Geovation awards programme 2009–2010. challenge.geovation.org.uk. Available at: https://challenge.geovation.org.uk/a/pages/custom-page-5 Accessed 6 May 2010

Parker CJ, May AJ, Mitchell V (2012a). The role Of VGI and PGI in supporting outdoor activities. Appl Ergon 44(6):886–894. Available at: http://www.sciencedirect.com/science/article/pii/S0003687012000816

Parker CJ, May AJ, Mitchell V (2012b) Understanding design with VGI using an information relevance framework. Trans GIS 16(4):545–560. Available at: http://onlinelibrary.wiley.com/doi/10.1111/j.1467-9671.2012.01302.x/full

Preece J et al (2002) Interaction design: beyond human-computer interaction. Wiley, New York

Richins ML, Bloch PH (1986) After the new wears off: the temporal context of product involvement. J Consum Res 13(2):280–285

Rieh SY (2002) Judgment of information quality and cognitive authority in the Web. J Am Soc Inform Sci Technol 53(2):145–161. Available at: http://onlinelibrary.wiley.com/doi/10.1002/asi.10017/full

Scharl A, Tochtermann K (2007) The geospatial web: how geobrowsers, social software and the Web 2.0 are shaping the network society, Springer, London

Schuett MA (1993) Information sources and risk recreation: the case of whitewater kayakers. J Park Recreat Adm 11(1):67–77. Available at: http://js.sagamorepub.com/jpra/article/view/1796

Spradley JP (1980) Participant observation. Holt, Rinehart & Winston, New York

Stake RE (1995) The art of case study. Sage Publications, Thousand Oaks

Sun G, Song W (2009) Using mobile GIS as volunteered GI provider. In: First IEEE international conference on information science and engineering (ICISE). Nanjing, China, pp 2229–2232

Tóth K, Tomas R (2011). Quality of geographic information—simple concept made complex by the context. In: Proceedings of the 25th international cartographic conference and the 15th general assembly of the International Cartographic Association. ICC, Palais des Congres, Paris, France

Tsou M-H, Yanow K (2010) Enhancing general education with geographic information science and spatial literacy. URISA J 22(2):45–54

Ummelen N (1997) Procedural and declarative information in software manuals. Rodopi B.V. Editions, Amsterdam

Wang P, Soergel D (1998) A cognitive model of document use during a research project. Study I. document selection. J Am SocInform Sci 49(2):115–133

Weiss AM, Heide JB (1993) The nature of organizational search in high technology markets. J Mark Res 30(2):220–233

Wilson P (1983) Second-hand knowledge: an inquiry into cognitive authority. Greenwood Press, Connecticut

Xiang Z, Gretzel U (2010) Role of social media in online travel information search. Tourism manage 31(2):179–188

Xiaolong G (2007) Practical research and development for embedded mobile GIS system based on PDA. Geospatial Inf 4:24–28

Yin RY (1994) Case study research: design and methods. Sage Publications, Thousand Oaks

Zeithaml VA (1988) Consumer perceptions of price, quality, and value: a means-end model and synthesis of evidence. J Mark 52(3):2–22

Zielstra D, Zipf A (2010) A comparative study of proprietary geodata and volunteered geographic information for Germany. In: Painho M, Santos MY, Pundt H (eds) Geospatial thinking: proceedings of the 13th AGILE international conference on geographic information science (AGILE). Guimarães, Portugal, pp 1–15

Chapter 5
Data Generation: VGI and PGI Data Sets

5.1 Introduction

In this book, research has sought to understand the way in which users perceive the utility of VGI to help aid them in their activities. The scoping study demonstrated that the users' decision to utilise VGI within *professional, personal* and *social* settings comes from their level of trust in the data and degree of homogeneity between the data user and the data contributor. More importantly, the scoping study suggested that the consumer would consider both VGI and PGI using the same criteria, in order to achieve their personal needs. Study Two highlighted how the consumer perceptions of VGI and PGI are influenced by their use requirements, where it is more useful to consider the attributes of the data (e.g. its currency) rather than the professionalism of the contributor. Study Two also demonstrated that the user judgement of trust is a key perception in the analysis of information during an information search, alongside cognitive authority and overall quality.

Research in the fields of quality (David and Jason 2008), *human–computer interaction* (Fogg and Tseng 1999) and *geo-sciences* (Idris et al. 2011) have highlighted *trust* and *credibility* as major factors in user judgements of online information. In the wider sense, Flanagin and Metzger (2008) highlighted the concerns for utilising VGI alongside PGI in terms of its *quality, reliability* and overall ability to add *value* to the user; situated as *credibility*. However, of most importance is the direct need to research the impact of such issues on the user. Although research within Studies One and Two have begun to outline the ways in which VGI as a single data source is perceived, how these perceptions influence the overall usability of a mashup is currently unknown. Consequently, the interaction between the various aspects in the user judgement relating to mashups containing VGI and PGI need to be understood.

5.2 Research Aims

The general aim of this chapter is to generate a VGI data set to be used to address the research aims above. Consequently, this study aims to address the specific objectives below:

1. Generate a body of VGI that can provide unique insights not presented through traditional PGI.
2. Combine VGI into a series of mashups that allow for integration in various websites, and can be used as the basis for controlled experimental study.

5.3 Study Rational

5.3.1 Selection of a Study Community

To address the study aims a set of participants was required whose information use would allow for an in depth exploration of how different forms of information are utilised, and how this influences their activities. In addition, participants were required to be already familiar with using both VGI and PGI relating to location-based information. This was necessary to ensure the outcomes of the study are applicable to *realistic* information use; rather than reactionary opinion of *first time use* (Baum et al. 1981). Additionally, the research which the user group undertakes before their activity must be understood as having a real and beneficial impact on future events for the user.

Previously research into the benefit of VGI within an end user context has been successfully conducted with:

- Parents pushing children in prams around an urban environment (Holone et al. 2007)
- Wheelchair users navigating an unfamiliar urban environment (Beale et al. 2006; Holone et al. 2008)
- Travellers with visual impairments navigating an urban environment (Kulyukin et al. 2008).

Ray and Ryder (2003) pointed out how even the most outgoing and *risk-taking* of the wheelchair user community actively and carefully evaluate the risks before travelling and engaging in travel. Importantly, within a travel context this is not experienced in the same fashion by able-bodied persons. This level of risk management as a central part of the group's activities allows for an enhanced connection between this investigation and the previous studies of this book.

5.3 Study Rational

Access and the ability for VGI to offer a reduced risk while engaging in travel situations has been a key theme in contemporary research. Therefore it was decided that wheelchair users (non-sensory or cognitively disabled) in travel situations was the most appropriate user group.

5.3.2 Selection of a Geographic Location for Research

London was chosen as the location of the investigation because of:

- Well established network of underground trains, buses and pedestrian routes allows for diverse travel scenarios to be presented to the participants.
- Large volumes of professional and volunteered information relating to the city and its travel network.
- Large and diverse number of locations, allowing travel from and to locations off the tourist map, which the participant is less likely to have first-hand experience of.

5.3.3 Selection of Travel Routes

The transport routes were restricted to those navigable for disabled travellers within a timeframe of 2–5 h (start to finish). In order to produce a representative description of the issues faced by travellers in London, it was also important to incorporate as many different transport modes as possible; train, underground; bus; light rail. Considering these factors, the following routes were selected (Fig. 5.1):

- *London Victoria* to Stratford via London Waterloo (bus, underground)
- *Stratford* to Angel Islington via Bow Street (bus)
- *Angel Islington* to Greenwich via London Bridge (bus, train)
- *Greenwich* to London Bridge (DLR light rail).

5.3.4 Selection of the Mashup Base Map

Since the focus of this book is the interaction between the user and the information presented to them, the role of the base map within the mashup was coincidental, being the relation of points of information to each other geographically (Crone 1968). Consequently, the map needed to be simple and neutral, so as not to overshadow the information presented within the mashup. After considering numerous maps and map styles (e.g. Bing, Google, Ordnance Survey), the CloudMade Pale Dawn map (CloudMade 2011) was selected for its appropriate simplicity; see Fig. 5.1.

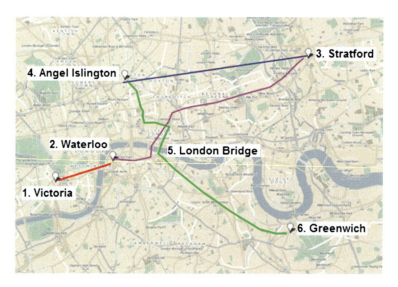

Fig. 5.1 Transport routes for VGI data collection

5.4 Investigation Overview

This study was an inductive investigation to produce two usable sets of information from volunteer and professional sources that may be utilised during phase two of this investigation (Study Three). A navigation route around London was selected, with both VGI and PGI data relating to transport accessibility issues collected. This was done through a combination of literature review and participant observation. During participatory observation, five wheelchair users travelled around the research route, accompanied by the researcher who took notes relating to their experiences and feelings about access issues, verbalised by the participant. Once collected, the VGI and PGI data sets were combined and displayed within a mashup.

5.5 Part A: VGI Data

5.5.1 Methods

5.5.1.1 Participant Sampling

In investigating the link between the number of VGI editors and the quality of the contributed project, Haklay et al. (2010) concluded that the first five contributors of VGI provide the bulk of accurate data, while successive contributions serve to

5.5 Part A: VGI Data

increase *accuracy* and *quality*. Although this may appear to be a relatively small figure, Holone et al. (2007) demonstrated a relatively small number of amateur volunteer contributions can be sufficient to generate good bespoke information relating to access and accessibility needs. Considering this, five participants were targeted.

For the purposes of this study, eligible participants were defined as:

- Physical disability necessitating the use of a wheelchair
- Only exhibits physical movement disabilities, excluding cognitive, sensory and audible disabilities
- Compatible with non-vulnerable persons description under the Loughborough University Ethics committee
- Confident in attempting travel via public transport.

Participants were recruited through a combination of social networking (e.g. twitter, Facebook, forums, etc.) and professional contacts with disability groups; i.e. NHS, Backup Trust, etc. All reasonable expenses encountered during the day were paid for by Loughborough University. Additionally, each participant was entered into a lottery for £150, drawn at the end of Study Three. The breakdown of participants involved within this study is presented within Table 5.1.

5.5.1.2 Data Collection

In studies into collecting VGI describing the built environment, various authors (Abley and Hill 2005; Cinderby et al. 2006; Evans 2009) demonstrated how the data collection method of *map walks* was effective, simple and insightful. Here, the participant is accompanied around the environment by the researcher, having their thoughts and opinions collected on route relative to their location. Consequently, this form of *Participatory Observation* was selected as the data collection method for this study. Due to the researcher being *able bodied* and unable to fully appreciate the level of severity access issues from the perspective of a wheelchair user, the position of *Observer-as-Participant* was taken. The structure of the data collection sessions was based on the principles of accessibility; see Table 5.2 based on Handy and Niemeir (1997).

Due to the level of difficulty that potentially faced the participants travelling on public transport (Options for Independent Living 2010), only one participant was involved in the study per day.

Table 5.1 A breakdown of the study participants by gender and wheelchair type

Chair type	Gender	
	Male	Female
Manual chair	2	1
Powered chair	2	0
Able bodied assistant	0	1

Table 5.2 Investigation of accessibility through data collection

Attribute of accessibility	How investigated through data collection
Spatial distribution of potential destinations	All points along the travel routes are accessed sequentially, allowing the cumulative effects of spatial distribution to be reflected in participant comments and opinions
The ease of reaching each destination	Participants asked to comments on situations related to accessing, travelling on and departing from the various transport modes along the travel routes
Magnitude, quality, and character of the activities	To every access issue commented on by the participant, they must also give an indication of how severe the issue is to their movement within the given environment

5.5.1.3 Procedure

During participant observation to produce the VGI data set, the following procedure was followed:

- Prior to data collection participants were provided with materials by email outlining the purpose of the data collection, procedure for the session, map of the travel route and terms and conditions of participation.
- Participants arranged to meet at London Victoria Station at a time and date that suited them, being the first point on the study's travel route.
- The participant and researcher set out along a pre-specified route (see Sect. 5.3.3) with the researcher guiding the choice of transport. Due to the physical limitations of the participants, and for their general ease, the researcher carried the *data capture sheet* to record the location, access issue and its severity During the observation period, prompts and questions were asked of the participant at relevant moments, such as *how did you find that?* or *after getting onto that underground train, is there anything that another wheelchair user should know before they arrive?*
- The participants were not told what to record, only that they should notify the researcher of all positive and negative accessibility issues that they see as important to another wheelchair user's making the same journey as them on a different day.

5.5.1.4 Analysis

Experiences and access issues encountered along the route were collected together to show the experience of all participants. The weighted mean was used to understand the average severity of the access issues identified by participants. Rather than present all issue severity scores collected through participation, average severity was included within the mashup, giving context to the collected VGI.

5.5.2 Results and Analysis

Figure 5.2 shows important stages of data collection with the study's participants.

With an attributed severity of five, the worst problems experienced were the noise of the train (London Waterloo underground station) and lack of information (Greenwich train station). Other serious problems (average severity of 4) related to the position, inclination and availability of ramps (Stratford bus station, Bow Church bus stop and London Bridge bus station) and poor information delivery (on bus, Stratford). Problems reported to be moderately severe (2.5–3 on the scale) included architectural barriers, such as gaps, steep inclines or curbs, and the absence of a wheelchair area (London Bridge train).

5.6 Part B: PGI Data

5.6.1 Methods

5.6.1.1 Data Collection

In order for the professional information to be applicable to the research aims it had to conform to the following specification:

- Structured geographic information produced by trained personnel (Fonseca & Sheth 2002)

Fig. 5.2 Data collection with participants during access surveys

- Provide detailed geographic information that can be verified and integrated at the national level (Goodchild 2007)
- Carry a degree of professional authority; i.e. be from an official body (Coleman et al. 2009).

PGI data were collected from the most widely accessed professional sources relating to wheelchair travel in London in order to give a comprehensive overview of the information currently available:

- *Direct Enquiries*: online repository of professional information about disabled access to a wide variety of locations around the UK (directenquiries.com 2011).
- *London Underground Step-Free Tube Guide:* available for pick up at all public transport locations in London (TFL 2010).
- *London Transport for London Website:* official information on all forms of public transport in London (TFL 2011d).

5.6.1.2 Procedure

Key literature (as identified in the background research) was searched, for information relative to the issue of wheelchair access at locations along the travel routes. Professional tourist information organisations such as Transport for London were also contacted to ensure that all information sources easily accessible by untrained persons were captured.

5.6.1.3 Analysis

Since PGI was gathered from existing professionally produced documents, the most direct and appropriate analysis technique was *content analysis*; described by Krippendorff (1980) as *a research technique for making replicable and valid inferences from data to their context.* Here the categories of general and specific geographic location—as identified through the travel routes and observation methods—provide a general framework for data collection. PGI sources were then searched for their applicability to the relative transport methods used through the journeys and the relative modes of travel. The collected data was then collated into a table, generating a *coded* summary of accessibility information for this study.

5.6.2 Results and Analysis

The professional data relating to wheelchair travel in London is presented in Table 5.3.

5.6 Part B: PGI Data

Table 5.3 PGI data relating to London travel wheelchair accessibility; specific locations

Location	Transport mode	Describe the access	Source
London victoria	Underground	No access to the underground for wheelchairs	TFL (2010)
	Train station	Staff on hand to help 24 h a day, 7 days a week, wheelchairs always permitted Train access ramps available, best booked at least 24 h in advance Step free access through the station; not to the underground	Network Rail (2011b)
Waterloo	Underground	Step between platform and the train = 50 mm Gap Between platform and the train = 70 mm	TFL (2010)
	Train station	Wheelchair access to the train and staff help to be confirmed by station operator Train access ramps available, best booked at least 24 h in advance Step free access through the station	Network Rail (2011c)
Stratford	Underground	Step between platform and the train = 50 mm Gap Between platform and the train = 78–85 mm	TFL (2010)
Angel Islington station	Underground	No access to the underground for wheelchairs	TFL (2010)
London bridge station	Train station	Staff on hand to help 04:00–01:00, 7 days a week, wheelchairs always permitted Train access ramps available, best booked at least 48 h in advance Step free access through the station	Network Rail (2011a)
Greenwich	Train station	Staff help Monday-Friday 06:00–21:30, Saturday 06:00–21:30, Sunday 06:00–21:30 Station is step free Train access ramps available, ask staff Wheelchair access to be confirmed by station operator	Southeastern (2011)

It should be noted that although bus routes play an important part in the travel routes within this study, no professional information was available regarding the bus stops or the area around the bus stops outside that detailed in Table 5.4.

5.7 Mashups

Once VGI and PGI data sets were collected, they were combined and presented using the UMapper mashup platform (www.umapper.com). This was done by creating icons on the map, which when clicked on would display the information collected during the study relating to that location; see Fig. 5.3.

Figure 5.4 demonstrates an overview of the base map as a user may see it, and the additional overlaid PGI information.

Table 5.4 PGI data relating to London travel wheelchair accessibility; general transport information

Transport mode	Describe the access	Source
Underground	Occasionally, a lift or escalator may be out of service. You can check this before you travel by using Journey Planner or calling our Customer Service Centre	TFL (2011c)
	You can ask a member of staff to help you get to the platform. All our staff have regular training on how to assist disabled passengers and will help you as far as it is safe to do so	
	Many stations have a vertical step into the train which may be as high as 12 inches (300 mm). There may also be a gap between the train and the platform. Please check if you can manage this before you travel. The Step-free Tube guide shows the step and gap at each step-free station	
Train	There is likely to be a step of a few inches between the platform and the train	TFL (2011b)
	We recommend that passengers requiring assistance give at least 24 h' notice by calling the helpline number below	
Bus	All of London's 8,000 buses are now low-floor vehicles (excluding Heritage buses on routes 9 and 15)	London and Partners (2011)
	Low-floor buses enable all customers, including people using wheelchairs to get on and off easily. Every bus also has a retractable ramp, which must be in full working order at all times	
	On all buses, there is room for one person using a wheelchair. Wheelchairs can be accommodated up to a size of 70 cm wide by 120 cm long. Wheelchair users have priority over everyone else for use of the wheelchair space. There is no limit on the number of assistance dogs allowed on the bus, as long as there is space	
	The wheelchair space on buses cannot take a wheelchair bigger than 70 cm in width and 120 cm in length	TFL (2011a)
	Each bus has a retractable ramp which makes access easier. Most wheelchairs, including motorised types, will fit onto buses but motorised scooters with handlebars can't be carried onto buses	
	If you are unable to board a bus because of a broken ramp, please wait for the next one and tell Customer Services as soon as possible on 0845 300 7000	

5.8 Discussion

Fig. 5.3 Creating a mashup with VGI and PGI Data (*left* a VGI node, *right* a PGI node)[2]

5.8 Discussion

The primary aim of the work in this chapter was to generate content for the study described in Chap. 8. However, it is interesting to reflect on the process of data generation and the production of map mashups. A short discussion of this therefore follows below.

5.8.1 Content of Collected Data

While the *impact* of the collected VGI on the user in comparison to the PGI data set is tested through Study Three, while PGI concentrated on objective facts and

---

[2] First published in Parker et al. (2012c)

Fig. 5.4 Example of mashup Set 1: PGI data investigation

practices, VGI focused primarily on the experiences of the user. This is in line with Goodchild (2010) who commented that PGI guarantees associated quality control, whereas VGI does not. In relation to this book, it is this level of professional quality control that prevents *emotional, personal* or *experiential* data from being presented within a PGI framework.

Further note should also be given to the way in which information is presented. While PGI revolves around a formal explanation of access features (e.g. step free, 50 mm gap, etc.) VGI presents access issues, often described in terms of the personal meaning and implication (e.g. gap too large, ramp too steep). On the basic level this may be the result of the untrained amateur describing a feature in the way which makes most sense to them rather than the trained professional delivering information in a tested, controlled and formal fashion (Goodchild 2008a; Tsou 2005; Tulloch 2008). This concept of PGI being objective and VGI being experiential is in line with the comments of van Excel and Dias (2011), who also noted that due to the limited levels of quality control associated with VGI, objectivity is a rarer occurrence than in PGI.

Considering the points above, the uniqueness of VGI is apparent, being a data collection method which captures the human centred issues. While this may be considered true at the current point, technologies such as HADRIAN (Porter et al. 2004) offer a professional and human centred way to assess the built environment. However, such systems are based on anthropometric data, and therefore cannot capture the emotional or experiential dimensions of the user in the environment. For example, current PGI data could be used to establish whether gaps between trains and platforms are too large, and future systems such as HADRIAN could be

5.8 Discussion

used to evaluate the suitability of the environment for the wheelchair user. However, neither method could ascertain data relating to the angle of bus ramps, concern over access at next stops or stress associated with waiting for access, or perceived treatment as a second-class citizen. Consequently, this study has demonstrated—to a degree—how VGI can capture information not traditionally covered by PGI, but also capture information which cannot be captured through PGI.

Within this study, the majority of access events provided by the participants related to negative experiences. This may be explained by Holone et al. (2008), who demonstrated that wheelchair users are more likely to contribute experiences about their environment when those issues relate to access issues faced at that moment in time. Consequently, sections along a travel route which did not result in access limitations were unlikely to be remarked on, even if their access *could* have been classified as good or excellent. The significance of this is that the possibility for single VGI projects to provide a *universal travel directory* as a *one-stop shop* for other homogenous users is limited due to the demonstrated focus on negative rather than all-encompassing experiences. However, considering the wider range of information available from both information types within this study, creating mashups utilising both VGI and PGI could create provide a more balanced perspective on the built environment.

5.8.2 Success of Data Collection

At the most basic level, this chapter succeeded in its aims of generating two data sets which could then be combined within a series of mashups for use in a further experiment. However, discussion needs to be given to the level of success and appropriateness with which those data sets were collected.

Although the data as collected through this study appears to represent the level of detail suitable for utilisation with effective and satisfying navigation products Holone et al. (2007, 2008) this is in the context of a low fidelity system. One of the most promising systems for assessing the human factors of the built environment is HADRIAN,[1] developed to assess key user requirements from access issues to information delivery (Porter et al. 2004). In investigating the role in which crowd sourcing may be utilised to enhance this system, Evans (2009) demonstrated that such amateur volunteer contributions can be effective and useful. However, data collected through this chapter did not reach saturation. There the degree of detail renders the data collected in this study inappropriate for use within the HADRIAN system. Therefore, it may be assumed that the data collected through this study is sufficient for delivering *informative* information about the built environment for other wheelchair users, but is probably not suitable for advanced *definitive* assessment.

---

[1] Build on the earlier AUNT-SUE project, described by Evans (2009).

A further consideration is the degree to which the VGI and PGI data agreed with each other. Because the participants in the study were not provided with a pre-specification of what to look for or comment on (e.g. pay special attention to kerb height) the two data sets are not directly comparable. This is despite both data sets focusing on the same geographic locations and for the same user group; wheelchair users. Therefore, a direct comparison is not feasible. However, from viewing both data sets it is clear that in some instances the VGI and PGI both comment on the same issue (e.g. London Victoria – no access to trains), while in others the VGI and PGI cover different issues (e.g. PGI: Greenwich station is step free, VGI: ramps are very steep). The picture which this creates therefore is not one of VGI confirming PGI, but PGI and VGI being used together to create a more complete image of the issues in the built environment.

This ability for VGI to add additional richness to the data set (where PGI is less complete) was originally presented by Goodchild (2007) who postulated the use of the world's six billion inhabitants as sensors to make up for this shortcoming in traditional GI. Additionally, utilising multiple information sources to counteract the limitations of a single information source (PGI or VGI) is supported by Hertzum et al. (2002) and Fallis (2004). Finally, Bishr and Janowicz (2010) also commented that as long as a proxy for establishing trust in VGI is put in place, the multiple combination of information has great potential for realising the concept of a fully integrated *digital earth*.

References

Abley S, Hill S (2005). Designing living streets: *a guide to creating lively, walkable neighbourhoods*, Transport research laboratory (TRL), London. Available at: http://trid.trb.org/view.aspx?id=769621

Baum A, Fisher JD, Solomon SK (1981) Type of information, familiarity, and the reduction of crowding stress. J Pers Soc Psychol 40(1):11–23

Beale L et al (2006) Mapping for wheelchair users: route navigation in urban spaces. Cartographic J 43(1):68–81

Bishr M, Janowicz K (2010) Can we trust Information?—The case of volunteered geographic information. In: Devaraju A et al (eds) Towards digital earth: search, discover and share geospatial data. Workshop at future internet symposium, Berlin

Cinderby S, Forrester J, Owen A (2006) A personal history of participatory geographic information systems in the UK context: successes and failures and their implications for good practice. In: L. McDowell (ed) Royal geographical society annual conference, RGS-IBG, London

Cloudmade (2011) Pale dawn, Cloudmade

Coleman DJ, Georgiadou Y, Labonte J (2009) Volunteered geographic information: the nature and motivation of producers. Int J Spat Data Infrastruct Res 4:332–358

Crone GR (1968) Maps and their makers: an introduction to the history of cartography, 4th edn. In: W. G. East (ed), Hutchinson, London

David R, Jason H (2008) Aesthetics and credibility in web site design. Inf process manage 44(1):386–399. Available at: http://www.sciencedirect.com/science/article/pii/S0306457307000568

References

Directenquiries.com (2011) Direct enquiries: the nationwide access register, Accessed 1 Aug 2011. Available at: http://www.directenquiries.com/

Evans G (2009) Accessibility, urban design and the whole journey environment. Built Environ 35(3):366–385

Fallis D (2004) On verifying the accuracy of information: philosophical perspectives. Libr Trends 52(3):463–487

Flanagin AJ, Metzger MJ (2008) The credibility of volunteered geographic information. GeoJournal 72:137–148

Fogg BJ, Tseng H (1999) The elements of computer credibility. In: proceedings of the SIGCHI conference on human factors in computing systems: the CHI is the limit. ACM, Pittsburgh,pp 80–87. Available at: http://dl.acm.org/citation.cfm?id=303001

Fonseca F, Sheth A (2002) The geospatial semantic web, UCGIS, Leesburg. Available at: http://www.ucgis.org/priorities/research/2002researchPDF/shortterm/e_geosemantic_web.pdf. Accessed 30 Oct 2012

Goodchild MF (2007) Citizens as voluntary sensors: spatial data infrastructure in the world of web 2.0. Int J Spat Data Infrastruct Res 2:24–32

Goodchild MF (2008) Commentary: whither VGI? GeoJournal 72(3):239–244

Goodchild MF (2010) Researching the geocrowd. Opening keynote. In: third spatial socio-cultural knowledge workshop. Shrivenham, Cranfield University, Cranfield

Haklay M, Basiouka S et al (2010) How many volunteers does it take to map an area well? the validity of linus' Law to volunteered geographic information. Cartographic J 47(4):315–322

Handy SL, Niemeir DA (1997) Measuring accessibility, an exploration of issues and alternatives. Environ Planning A 29:1175–1194

Hertzum M et al (2002) Trust in information sources: seeking information from people, documents, and virtual agents. Interact Comput 14:575–599

Holone H, Misund G, Holmstedt H (2007). Users are doing it for themselves: pedestrian navigation with user generated content. In: Al-Begain K (ed) NGMAST 2007: International conference on next generation mobile applications, services and technologies. IEEE, Cardiff, pp 91–99. Available at: http://ieeexplore.ieee.org/stamp/stamp.jsp?tp=&arnumber=4343406

Holone H et al (2008) Aspects of personal navigation with collaborative user feedback. In: NordiCHI'08: Proceedings of the 5th nordic conference on human-computer interaction: building bridges. Lund/IKDC, NordiCHI, Sweden, p 182

Idris NH, Jackson MJ, Abrahart RJ (2011) Map mash-ups: what looks good must be good? In: Emma Jones C et al (eds) Proceedings of the 19th GIS research UK annual conference. GIS Research UK, Portsmouth, p 119

Krippendorff K (1980) Content analysis: an introduction to its methodology. Sage, New York

Kulyukin V et al (2008) The blind leading the blind: toward collaborative online route information management by individuals with visual impairments. In: Proceedings from the association for the advancement of artificial intelligence (AAAI) spring symposium, Stanford University, California

London & Partners (2011) London buses, visitlondon.com. Available at: http://www.visitlondon.com/travel/getting_around/london-bus. Accessed 7 Aug 2011

Network Rail (2011a) London Bridge, nationalrail.co.uk. Available at: http://www.nationalrail.co.uk/stations/lbg/details.html. Accessed 11 Aug 2011

Network Rail (2011b) London Victoria, nationalrail.co.uk. Available at: http://www.nationalrail.co.uk/stations/vic/details.html Accessed 11 Aug 2011

Network Rail (2011c) London Waterloo, nationalrail.co.uk. Available at: http://www.nationalrail.co.uk/stations/wat/details.html Accessed 11 Aug 2011

Options for independent living (2010) Using public transport with choice, control and confidence. Essex County Council, Chelmsford

Parker CJ, May AJ, Mitchell V (2012c) Using VGI to enhance user judgements of quality and authority. In: Whyatt D, Rowlingson B (eds) proceedings of GIS research UK 20th annual conference. GIS Research UK, Lancaster, pp 171–178

Porter JM et al (2004) Beyond jack and Jill: designing for individuals using HADRIAN. Int J Ind Ergon 33(3):249–264

Ray NM, Ryder ME (2003) 'Ebilities' tourism: an exploratory discussion of the travel needs and motivation of the mobility-disabled. Tourism Manage 24:57–72

Southeastern (2011) Greenwich, nationalrail.co.uk. Available at: http://www.nationalrail.co.uk/stations/gnw/details.html. Accessed 11 Aug 2011

TFL (2010) Step-Free Tube Guide, tfl.gov.uk. Und, London. Available at: http://www.tfl.gov.uk/assets/downloads/step-free-tube-guide-map.pdf

TFL (2011a) Transport accessibility: buses, tfl.gov.uk. Available at: http://www.tfl.gov.uk/gettingaround/transportaccessibility/1171.aspx. Accessed 5 Aug 2011

TFL (2011b) Transport accessibility: rail, tfl.gov.uk. Available at: http://www.tfl.gov.uk/gettingaround/transportaccessibility/1175.aspx. Accessed 5 Aug 2011

TFL (2011c) Transport accessibility: tube, tfl.gov.uk. Available at: http://www.tfl.gov.uk/gettingaround/transportaccessibility/1169.aspx. Accessed 5 Aug 2011

TFL (2011d) Transport for London, tfl.gov.uk. Available at: http://www.tfl.gov.uk/. Accessed 5 Aug 2011

Tsou M-H (2005) An intelligent software agent architecture for distributed cartography knowledge bases and internet mapping services. In: Peterson MM (ed) Maps and the internet. Elsevier Ltd., Kidlington, pp 229–243

Tulloch DL (2008) Is VGI participation? from vernal pools to video games. GeoJournal 72:161–171

Van Excel M, Dias E (2011) Towards a methodology for trust stratification in VGI. In: VGI pre-conference at association of american geographers (AAG), Seattle, Washington, pp 1–4

Chapter 6
Study Three: Assessing the Impact of VGI

6.1 Introduction

Study Two demonstrated that in a *realistic* use scenario, consumers are more likely to use VGI and PGI alongside each other (where available) in order to converge on a *truth* than to use individual VGI or PGI data sets. However, as highlighted by Rieh (2002) the way in which information is perceived by a consumer during an information search is based on a multitude of influences. The perception of information is critical, since it will influence the extent to which it is used.

Before this chapter, the data generation chapter focused on the generation of a VGI and a PGI data set, both describing the same geographic region so they may be compared and contrasted. The aim of the data generation chapter was primarily to generate the data sets necessary for the study described in this chapter. For this study it is important to focus on a consumer user group that is the same as the contributor group since Studies One and Two highlighted this form of homogeneity to be both common and beneficial in neogeography. Through the manipulation of variables in the information presentation, this study seeks to understand the unique abilities for VGI to influence the user perceptions when combined with PGI through an online interactive mashup.

This focus on user perceptions is critical to fulfilling the current need for design guidance on mashup creation (Idris et al. 2011a). Several authors (Boin and Hunter 2006; Devillers et al. 2002; Frank 1998) have commented that assessment by the descriptors of information alone (its metadata) is difficult and potentially inappropriate, while eliciting user feedback has been proposed as a useful and effective way of assessing the quality and appropriateness of online information (Comber et al. 2007).

The study reported within this chapter was an empirical investigation into the extent that including VGI alongside PGI, or including *and telling the user* that there exists VGI alongside PGI within a mashup, influences the user experience of a neogeographic system. In particular, this study focused on the effects on the *trust* that users place in information. Study Two highlighted how trust—both in information and as an emergent property to utilising information—is a critical factor in the user's evaluation of VGI. Since *trust* may be taken as *confidence in or reliance on some quality or attribute of a person or thing, or the truth of a statement* (OUP 1989), this study aims to derive it from the user's perceptions of *quality* and *authority* of a mashup. If including VGI within such information portals increases the user perception of neogeography then this study would demonstrate some of the effective boundaries and influences of VGI from a human factors perspective. Importantly, this experiment was based on the perceptions of information by users, rather than objective and repeatable measures of *truth*.

6.2 Research Aims

The research aims for study three are:

1. The extent to which actually including VGI within the mashup alongside PGI affects the users' judgements;
2. The extent to which the users react to the information that their mashups contain VGI;
3. The extent to which aspects of the users' judgements that may be harnessed to optimise the design of future mashups combining both VGI and PGI information.

6.3 Study Rational

As this study uses the data from the data generation chapter, the same study community (wheelchair users without cognitive or sensory disabilities in a travel context) was carried forward into this study.

6.4 Methodology

6.4.1 Overview

This study comprised an online experiment to assess the influence of presenting users with mashups containing PGI or PGI + VGI, and the influence of telling users that their mashups contain PGI or PGI + VGI has on their judgements of the websites quality and authority. Four independent groups of participants were used, each with a unique combination of the independent variables. The mashups as presented to the participants contained information on public transport around set routes in London, comprising bus, overground, underground and light rail trains. Participants were asked to consider how confident and comfortable they would be making the journeys as presented to them in the near future if the only information they had was that within the mashup. A Likert scale questionnaire was then presented to the participants in order to collect their judgements, including the dependant variables of *quality* and *authority*.

6.4.2 Experimental Variables

6.4.2.1 Independent Variables

Within this study, the independent variables were as:

1. Information as presented to the participant
   a. Mashup only contains PGI
   b. Mashup only contains PGI + VGI
2. Information as told to the participant
   a. Participant told that their mashup contains PGI
   b. Participant told that their mashup contains PGI + VGI.

Due to time and budget constraints of the study, *VGI on its own* was not included within the independent variables. This was because doing so would vastly reduce the likelihood of achieving the minimum numbers of participants required by the assumptions of the statistical tests. While reducing the number of conditions (and thus groups) increases the number of participants per group within the study, the main reason for this decision was that this study investigated the influence of VGI on PGI, rather than to understand the differences between VGI and PGI.

6.4.2.2 Dependant Variables

The dependant variables within this experiment needed to be dimensions of user judgement that have been demonstrated to be related to holistic perceptions of information within an online context. In investigating the judgement of information involved in an interaction by a user, Rieh (2002) presented a model to describe how users perceive *quality* and *cognitive authority* in online information; see Fig. 2.8, page 89. These judgements are *good, accurate, useful, important, trustworthy, credible, reliable, scholarly, official* and *authoritative*; see Table 2.11, page 90. This framework has also been used in a similar and recent study by Idris et al. (2011a), giving additional demonstrated credibility to its appropriateness.

Consequently, the dimensions of information judgement as highlighted above make up the dependant variables of this study.

6.4.3 Experimental Design

Users were presented with a mashup unique to their assigned group according to the independent variables; see Table 6.1.

The participants were presented with a number of travel routes (see Sect. 6.4.5) that create engagement between the participant and the information. They were then asked to consider how they would feel making that journey tomorrow if the only information they had was that presented to them. This allowed judgements relating to the mashup as a whole to be formed. A similar approach was successfully undertaken by Collins (2006) who presented a data set online to experiment participants while informing them that it was either from source A or B in order to understand perceived bias in information judgement perceptions. Previous research has shown such an approach to be highly relevant and beneficial when researching GI *use* and *utilisation* (Bishr and Mantelas 2008; Idris et al. 2011b; Mummidi and Krumm 2008).

6.4.4 Design of the User Judgement Survey

6.4.4.1 Likert Scale Questionnaires

The questionnaire provided to the participants at the end of the experiment was designed to investigate the influence of the independent variables on the dependant variables. Consequently, the structure of the questionnaire was set to reflect the structure of Rieh's facets of judgements—quality and authority.

Because the evaluative judgements made by the user on the information comprised their *opinions, attitudes* and *beliefs* (Albaum 1997; Mizumoto and Takeuchi 2009) the most appropriate method of investigating the participant response to information presented in the study was through *Likert Scales*. In forming the statements within the Likert Scale, words and phrases used by the participants in the work of Rieh and Belkin (2000) to relate to the facets of

Table 6.1 Group conditions by the two variables, what the map contained (PGI or PGI + VGI) and what the participants were told the map contained (PGI or PGI + VGI)

		Information presented in Mashup	
		PGI	PGI + VGI
Participant told what Mashup contained	PGI	Group 1	Group 3
	PGI + VGI	Group 2	Group 4

6.4 Methodology

information judgement were utilised. However, the category relating to the scholarly nature of the work was removed from the survey since it held no direct relevance to the investigation within the study.

Tables 6.2 and 6.3 contain the questions as presented to the participants within the questionnaire. For a full overview of the arrangement of the questions as presented to the study participants. Following the advice of Levine et al. (1993), each section of sub section of the question sheet aimed to provide 50 % positive and 50 % negative statements to the participant.

6.4.4.2 Validity and Reliability

The wording, structure and presentation of the Likert scale was tested within the pre-pilot and pilot stage of the website analysis; see Sect. 6.4.5.2. *Factor analysis* was required within the study to ensure that the data collected presents a faithful measure of the factors being investigated. Since the pre-test and pilot showed favourable reactions towards the survey (as presented within Sect. 6.4.4, page 98), *confirmatory factor analysis* was run post-data collection to ensure suitably robust sample sizes per factor could be reached.

Table 6.2 Questions on the judgements of information quality (based on Rieh and Belkin 2000)

Values	Likert scale statements (1—completely disagree, 5—completely agree)	
Good	The information provided by the maps	Did a good job at informing me about accessibility
		May not have been the best possible
		Was for my needs perfect
		Could have been better
Accurate	The content of the maps	Was as accurate as I could hope for
		Was not always correct
		Should be considered right
		Was not always as precise as I would want it to be
Current	The materials I engaged with on the maps	Reflected the current conditions well
		Seemed to be old and out of date
		Appeared to have been generated recently
		Did not capture the timely importance of travel information
Useful	Overall, I found the maps	Useful for my needs
		Useless for what I needed to find out
		Informative in its contents
		Did not help me feel confident I could travel without problems
Important	The data presented to me through the maps	Would be important to me when planning future travels
		Would be unimportant to me when planning future journeys
		Does not need to include any more information
		I would require more diverse information

Table 6.3 Questions on cognition of information authority, (based on Rieh and Belkin 2000)

Values	Likert scale statement	
Trustworthy	After using the website	I do not believe it would help me travel without access issues
		I feel I can rely on the information to help me travel freely
		I do not have faith in the quality of the content
		I feel confident that the information provided is true
Credible	I feel like the information provided	Was credible
		Did not provide information from sources that were experienced in disabled travel
		Came from sources that knew that were knowledgeable
		Did not come from credible sauces
Reliable	I feel I	Can rely on the information to help me travel without encountering access issues
		May need other forms of information to help me travel freely
		Can depend on the information when I go travelling
		Would rather use other forms of information when planning a trip
Official	The maps should be considered	As presenting official information
		As secondary to official websites
		Worthy of inclusion on key tourist websites
		As containing unofficial information
Authoritative	The information I was presented with	Felt authoritative
		Is not respected in my mind
		Should be considered worthy of respect
		Did not feel like it embodied much authority

6.4.5 Design of the Website

6.4.5.1 Initial Development

Both PGI and VGI data presented to the participants through the experimental mashups was collected and collated prior to the planning and execution of this study. For full details, see Chap. 5. An example of the mashups developed within mashups is seen in Fig. 6.1.

As participating within the experiment was voluntarily undertaken in a home setting, it was necessary to keep participants engaged during their time on the website to prevent the participants leaving the session prematurely. The website was therefore produce in accordance with the experiment website guidelines of Frick et al. (2001) and Reips (1996, 1999):

- Make web pages shorter and more attractive the further participants get.
- The loading time at the start of the website should be short in order to engage participants with low interest or little time.

6.4 Methodology

Fig. 6.1 Example of Mashup set 2: PGI + VGI data (Parker 2011)

- Announce a lottery with prizes for all successful participants.

The website as developed within the experiment is presented in Fig. 6.2.

As detailed in Table 6.1, various levels of information are required to be presented to the user. While a number of delivery methods are available, instructional videos hosted online (YouTube) were felt to be the most appropriate since they:

- Ensures consistent delivery of information to all participants.
- May provide a level of professionalism in the instruction, increasing the cognitive authority of the website equally for all groups as not to introduce an experimental variable.
- Provide an engaging experience which is complimentary to the interactive nature of Web 2.0 and VGI (Bishr and Mantelas 2008; O'Reilly 2005).
- Allow simple dissemination of information over the internet 24 h a day without requiring the researcher to be present.

As pointed out by Rieh (2002) the participants need to be presented with active information-seeking tasks in order for them to form valid judgements, and thus allow the investigation to gain a true understanding of how information is used and perceived in a realistic situation. The tasks within the experiment were characterized as *generic tasks* in order to outline the information seeking activity, but not to restrict the specific experiences. This left the perception of the information unconstrained by the experiment. In order for the participants to encounter *problems* on their virtual journey, routes were selected for the initial data gathering which went through known problem spots. These were identified through the

Fig. 6.2 Example of Mashup presenting VGI alongside PGI[1]

Transport For London website a few days before the VGI data was gathered. At each of the tasks, the participants were instructed to consider:

- Previous journeys made which may be similar (e.g. train travel)
- What information they would need if they were to make a similar journey
- To what extent the information presented to them fulfils their information needs (e.g. completely, partly, not at all)
- How confident the information would make them feel if they were to conduct that journey in the near future.

6.4.5.2 Validity and Reliability: Pre and Pilot Testing

A critical element of the validity of this experiment is the choice to host the experiment online, therefore accessible by the participants in their natural home environment rather than within a laboratory. In practice, participants would visit the experiment website via their home computer at a time of their choosing, be allocated to a group, given a pre-experiment briefing via video and then presented with a series of mashups before answering a survey. During use of the mashups, participants were asked to consider the route shown and how they would feel making that journey tomorrow if that was all the information they had. This, as

---

[1] First published in Parker et al. (2012).

6.4 Methodology

Weathington et al. (2010, p. 269) pointed out, would influence subjects to *"do what they think the researcher wants them to do"* rather than what they would do in their natural setting. Therefore by presenting the information to participants in their *realistic* natural setting they are less likely to engage in this form of *game playing* and the factors being investigated by the experiment are more likely to be those actually measured.

A pre-test was run exploring the initial mechanisms of the study using non-disabled members of Loughborough Design School. The aim of this *pre-test* process was to trial some or all aspects of the instrument to ensure there are no unanticipated difficulties (Alreck and Settle 1995). This consisted of a custom, interactive website, embedded instructional videos, embedded interactive mashup and full survey. Critically, the independent variables presented to the participants during the pre-test were (1) the mashup contained only PGI and (2) participants were told that the mashup contained PGI + VGI. The website may be viewed at http://chris210.wix.com/free-traveller. In total eight participants took part in the pre-test. The key outcomes from the pre-test were:

- General website usability improvements needed to convey a *high quality* user experience through the experiment.
- The need for more demonstrative, clearer and professional instructional videos
- The need for clearer presentation of information within the mashup.

Additionally, protocol analysis was conducted to assess the suitability of the experiment survey, see Sect. 6.4.4; page 98. Here, individual participants within the pre-test group were asked for their opinions on the survey questions; notably what the intention of the questions were and what was being asked (Ericsson and Simon 1993). Within the survey, minor issues relating to grammar and clarity were corrected. Overall, the survey was found to be suitable and appropriate by the pre-test protocol analysis.

Once the website, mashups and survey elements of the experiment had been created and adjusted according to the pre-test, a small scale pilot study was conducted to ensure that the experiment would run as designed. In total 36 wheelchair users (17 male, 19 female) engaged with the interactive survey, providing data for the study using the online website. From feedback collected from participants it was clear that although the mashup and survey was appropriate and effective, the website needed to have better usability in order to prevent some users from abandoning the interactive survey part way through due to frustration. The outcomes from the pilot test and how they were addressed is presented in Table 6.4.

6.4.5.3 Website Usability Assessment

A usability assessment of the experiment website was conducted to ensure that the judgements as measured by the dependable variables were the result of the independent variables, rather than overly influenced by a poorly designed website.

Table 6.4 Pilot test issues and how they were addressed

Usability issue	How addressed
Apple Mac users experienced 'load' problems with the website	Help section for Apple Mac users added explaining how to fix runtime issues
Some users found navigating the maps difficult	A help section for navigating the maps, was added to the tutorial practice map pages
Some participants were unsure of exactly what to do on the maps	Simple text added to the map mashup pages explaining that all that was required was considering the Information
The end of the survey seemed uncertain	A video message was added to the end of the survey thanking participants for their time and asking them to share the survey with others

This was achieved by including questions based on the Software Acceptance Questionnaire (Maguire 1998) within the Likert Scale survey. The assessment found that although clarity of information delivery could be improved upon, the website exhibited of high-level overall usability; see Fig. 6.3.

Additionally, when a Two-Way MANOVA test was run, no statistically significant interactions were observed between the groups. The assessment found the website to have suitable usability for the function of the experiment.

6.4.6 Participant Sampling

6.4.6.1 Demographics Specification

It was important that the participants who engaged with the experiment were in a position to critically evaluate the information to form realistic judgements. In order for this to be achieved, the following screening criteria were generated.

- Physical disability which limits movement and necessitates the use of aids similar to and including wheelchairs.
- Only exhibits physical disabilities, excluding cognitive and sensory disabilities
- Compatible with non-vulnerable persons description under the Loughborough University Ethics committee, except in circumstances listed above.
- Full access to and competence using a PC, Laptop, Tablet or other internet enabled computer with a full sized screen; e.g. excluding pocket portable devices such as mobile phones.
- Have a good to excellent familiarity and confidence using online maps; e.g. Google Maps.

Due to the *virtual tour* nature of how information was delivered to the participants, no existing knowledge of London public transport was required of the participant.

6.4 Methodology

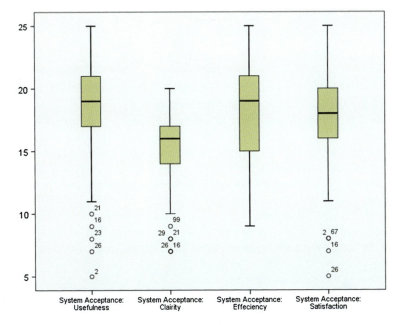

Fig. 6.3 Box plots representing user acceptance of the website

6.4.6.2 Sample Representativeness and Error

The issue of culture has been raised as a potential differentiator in holistic and *analytic perception, usability* and *cartographic perception* (Edsall 2007; Nisbett and Miyamoto 2005; Shi 2010). This study sought to contact participants within the international community, although limited to economically developed English speaking countries; i.e. United Kingdom, USA, Canada and Australia. This also increases the number of participants responding to the survey, making the outcomes more statistically reliable, as well as increasing the overall external validity of the results.

While sampling errors may be considered unavoidable (Caswell 1995) its effect on this research study is limited by using appropriate and diverse participant recruitment, accessing a wide spectrum of participants from within the wheelchair user community.

6.4.6.3 Recruitment

In order to access as wide a variety of participants as possible (in line with the boundary conditions) multiple points of contact were used; see below:
- Wheelchair specific disability services and groups
- Internet forums that served an international audience

- Social Media presence and adverts (i.e. Facebook, Twitter, Google +) targeted at residents of the target countries.

In order to accommodate the wide variety of locations where recruitment took place, the experiment was branded *Free Traveller*. This provided an effective synergy between multiple social media profiles (e.g. Twitter, Facebook, Google +, etc.), printed and electronic flyers and the website itself.

By hosting the experiment within an online and interactive website, the issue of participant availability was reduced, since they may take part in the experiment at any time they have access to the internet. Within the UK 30.1 million adults (60 % of the UK population) accessed the internet every day in the UK, and only 9.2 million did not ever access the internet (ONS 2010). Considering the wider use of the internet worldwide, Fox (2010) commented that 54 % of adults living with a disability use the internet in the USA, compared with 81 % of able-bodied adults (Carter 2011). Therefore, access to the internet was not a barrier to participation in the experiment. Overall, by hosting the experiment online the limitations on time, energy and resources posed by traditional experiments held within laboratory conditions are overcome, as well as increasing the ecological validity of the research.

In order to run the appropriate statistical tests for the analysis within this study, it was decided to achieve a representative sample of 100 participants, before ending data collection and running statistical analysis.

6.4.6.4 Rewarding Participant Time

In order to reduce the number of participants who drop out of the experiment part way through, and to maximise engagement with the website from first visit (Frick et al. 2001), a financial incentive of being entered into a lottery to win £150 was offered to participants who successfully completed the survey. Frick et al. (2001) demonstrated that providing incentives to participants in the form of a lottery reduced the number of dropouts of the online experiment yet did not provide a bias in the answers that they provided.

6.4.7 Procedure

Participants were contacted through a variety of methods, as detailed in Sect. 6.4.6.3, page 96. Following this, participants were directed to the Free Traveller website. They then worked through the following stages:

- *Stage 1*: Placing them in experiment groups
- *Stage 2*: Delivering basic instructions
- *Stage 3*: Telling them that their maps contained PGI or PGI + VGI
- *Stage 4*: Using the Mashup
- *Stage 5*: Assessment questionnaire
- *Stage 6*: Giving out Prize.

6.4 Methodology

None of those involved with the collection of the VGI data set took part in the online experiment to prevent contamination of data by experience.

6.4.8 Statistical Analysis

6.4.8.1 Overview

The first stage of analysis was *confirmatory factor analysis*, selected to ensure the data faithfully represents the factors being measured. To understand how the dependant variables are influenced by the independent variables within this experiment, an appropriate statistical method based on analysis of variance is required. Although the data created through using Likert Scale is ordinal, the most powerful tool was Two-Way Multivariate Analysis Of Variance (MANOVA). In this case, the *two-way* refers to the number of independent variables. Using MANOVA also reduces the risk of a *type 1* inflation error in the data analysis (Pallant 2010). MANOVA is primarily designed for parametric data, however, it may only be used with ordinal (non-parametric) data when all assumptions are met prior to its calculation; particularly Kolmogorov–Smirnov achieving significance and thus demonstrating sufficient normality within the data.

In the case where assumptions within the data set are violated, the non-parametric equivalent of MANOVA would be used; Kruskal–Wallis. However, this test is not as powerful as MANOVA in understanding the influence of the independent variables on the dependable variables (Caswell 1995; Field 2004; Pallant 2010) and thus is a backup approach rather than a main tool.

Prior to statistical analysis, the following considerations were given to the data:

- *Sampling error* within the analysis was reduced through (1) ensuring that enough participants were sampled to satisfy the assumptions for each statistical tool and (2) sampling users from geographically dispersed regions, embodying a ride range of mobility disabilities.
- *Measurement error* was reduced by collecting data through a specially designed survey, measuring only those factors of interest to the experiment on a five point Likert scale.
- *Estimation error* was reduced by ensuring that the data set met the assumptions required by the statistical tools (including outliers) analysis was conducted.

6.4.8.2 Confirmatory Factor Analysis

In order to assess the suitability of the grouping or utilisation of the various dependable variables (Keller 2006; Pallant 2010), confirmatory factor analysis was selected to ratify the outcomes of the survey.

The 10 dependant variables of the *User Judgement* Survey were subjected to Principal Components Analysis; PCA. Prior to performing PCA, the suitability of data for factor analysis was assessed. Inspection of the correlation matrix revealed the presence of many coefficients of 0.3 and above, demonstrating sufficient correlation (Pallant 2010). The Kaiser-Meyer-Oklin value was 0.93, exceeding the recommended value of 0.6 (Kaiser 1970, 1974) and Bartlett's (1954) Test of Sphericity reached statistical significance, supporting the factorability of the correlation matrix.

Principal components analysis revealed the presence of one component with an eigenvalue exceeding 1.0, explaining 71.2 % of the variance. An inspection of the screeplot revealed a clear break after the first component. Using Catell's (1966) scree test for rotation sums of squared loadings, it was decided to retain one component for further investigation. Parallel Analysis also showed only one component with an eigenvalue exceeded the corresponding criterion values for a randomly generated data matrix (Watkins 2000) of the same size (10 variables × 101 respondents).

6.4.8.3 Scale Reliability Measures

Since the *judgement* scale was developed from the work of Rieh (2002) specifically for this experiment no previous data are available on its internal consistency. In the current study, the Cronbach alpha coefficient was 0.95; suggesting exceptionally good internal consistency in the scale.

6.4.8.4 Descriptive Statistics

A breakdown of the participants involved within the study by gender is given below in Table 6.5 and by geographic location in Table 6.6.

A two-way between-group multivariate analysis was performed to investigate the influence of the confounds of *gender, country of residence, regional settlement type, computer use, confidence using online maps* and *confidence travelling* on the dependant variables of the experiment. Preliminary assumption testing was conducted to check for normality, linearity, univariate and multivariate outliers, homogeneity of variance–covariance matrices, and multicollinearity, with no

Table 6.5 Breakdown of participants per group by gender

Group	Gender		Total
	Male	Female	
1	11	12	23
2	16	17	33
3	6	16	22
4	7	16	23

6.4 Methodology

Table 6.6 Breakdown of participants by location

Country	Frequency
Australia	2
Canada	10
Ireland	2
New Zealand	3
UK	63
USA	20
Other	1
Total	101

serious violations noted. For all confounds, no statistically significant interactions were observed.

The confidence and familiarity of the user with online mashups was a potential limiting factor to the analysis. Negative judgements may be formed during the experiment not as an influence from the independent variables, but from the lack of confidence in using the system; a variable not covered by this investigation. However as Fig. 6.4 demonstrates, the vast majority of participants were very confident using online maps prior to engagement with the Free Traveller experiment. Consequently, the influence of participants being uncomfortable using mashups similar to those included in the experiment can be considered negligible.

6.5 Hypotheses

Based on previous research being applied to the pre-specified design model, the following null and alternative hypothesised were constructed; see Table 6.7. Due to the uncertain nature of the influence of VGI and PGI on users, a *2-tailed* hypothesis was taken.

6.6 Results and Analysis

6.6.1 Two-Way MANOVA

A two-way between-group multivariate analysis was performed to investigate (1) the inclusion of VGI alongside PGI within a mashup, and (2) the influence of being told a mashup contains VGI alongside PGI, on the user judgement of mashups quality and authority.

Preliminary assumption testing was conducted to check for normality, linearity, univariate and multivariate outliers, homogeneity of variance–covariance matrices, and multicollinearity, with no serious violations noted. Importantly, the

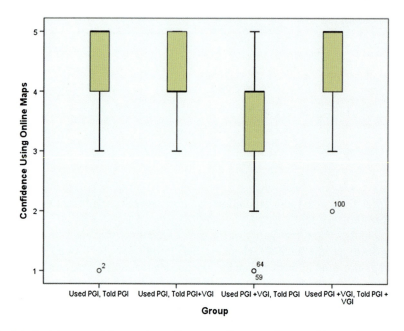

Fig. 6.4 Participant confidence using online maps (1 = very low confidence, 5 = very high confidence)[2]

Table 6.7 Alternative hypotheses within study three

Null hypothesis	Alternative hypothesis (2-Tailed)
No significant interactions in the quality and authority judgements of maps containing PGI or those containing PGI + VGI	Presenting groups with maps containing PGI + VGI (rather than just PGI) influences their judgements of quality and authority
No significant interactions in the quality and authority judgements of maps when users are told they contain PGI or PGI + VGI	Informing the participant that the information they are using is volunteer generated influences their judgements of quality and authority

Kolmogorov–Smirnov test for normality was significant, demonstrating the appropriateness of MANOVA as the statistical analysis tool. However, due to the statistically high correlation between the *authority* elements of *trustworthiness* and *reliability* ($\rho = 0.846$), the item *reliability* was removed from the data set since of the pair it exhibited the highest level of correlation with other items. This was done to meet the assumptions of MANOVA (Pallant 2010) and to allow insight into not only the statistical significance of dependable variables, but also their effect sizes (Field 2004).

---

[2] First published in Parker et al. (2012).

6.6 Results and Analysis

The Levene's Test for Quality of Variance produced a statistically significant outcome ($\rho = 0.04$). Consequently, it was necessary to use a lower alpha (0.025) to be sure of significance in the univariate F-test (Tabachnick and Fidell 2001).

Although the *N* values for the data were not equal, making *Pillai's Trace* the most appropriate multivariate test (Tabachnick and Fidell 2001; Pallant 2010), due to the small number of groups involved in the data, the F-tests for Wilks' Lambda, Hotelling's Trace and Pillai's Trace were identical. Therefore, Wilks' Lambda was used for its applicability to general use (Tabachnick and Fidell 2001).

No statistically significant interactions were observed between those groups who were told that their mashups contained PGI + VGI and those groups who were told that their mashups contained only PGI, $F(9, 89) = 1.20$ $\rho = 0.304$; Wilks' Lambda $= 0.89$; $\eta p^2 = 0.108$.

There were statistically significant interactions between those groups who were presented with mashups containing PGI + VGI and those groups who were presented with mashups containing only PGI on the combined dependant variables, $F(9, 89) = 3.91$, $\rho = 0.000$; Wilks' Lambda $= 0.72$; $\eta p^2 = 0.283$.

When the results for the dependent variables related to the information as presented to the participants were considered separately, the only user judgement to reach statistical significance, using a Bonferroni adjusted alpha level of 0.006,[3] was currency: $F(1, 97) = 10.81$, $\rho = 0.001$, $\eta p^2 = 0.10$. The ηp^2 of 0.10 represents 10 % of the variance in perceived currency scores explained by belief that the mashup in use contains VGI. Under the generally accepted criteria of Cohen (1988) this constitutes a medium effect size. An inspection of the mean scores indicated that those who believed that their mashup contained PGI + VGI reported slightly higher levels of perceived currency in the map date ($\bar{x} = 13.98$, $SD = 2.68$) than those who had believed that their mashup contained only PGI ($\bar{x} = 12.48$, $SD = 3.45$).

At no point was a statistically significant interaction between the fixed variables observed within this MANOVA test.

In assumption testing conducted prior to MANOVA, all of the dependable variables were demonstrated to have non-significance according to the Levene's Test. Therefore, non-significant outcomes from the MANOVA test may be discussed in terms of their *significance of similarity influence*; i.e. as opposed to significance of difference as measured by MANOVA and other analysis of variance tests.

6.6.2 Sample Size Estimation

Due to the low level significance of variables within the data set, it was useful to apply further inferential analysis to *predict* possible significance within larger

---

[3] Bonferroni adjustment (0.006) = experiment alpha (0.05)/number of comparisons (9).

groups. Sample size estimation[4] aimed to predict the number of participants that could be in future similar research. Figures were reached with the aid of *Appendix D* of Murphy and Myors (2004).

Table 6.8 demonstrates the estimated sample sizes required for achieving statistical significance using MANOVA for each of the dependant variables found insignificant within the experimental data set; N = 101.

Although estimation of sample size was increased to N = 606, no significance was predicted relating to accuracy or authority.

By estimating the sample size to be N = 202, significant interactions were predicted between those who were told that their mashups contained PGI + VGI and those who were told that their mashups contained only PGI.

Table 6.9 demonstrates the estimated sample sizes required for achieving statistical significance using MANOVA for each of the dependant variables found insignificant within the experimental data set.

Although estimation of sample size was increased to N = 606, no significance was indicated relating to importance or accuracy.

6.6.3 Testing of Independent Variables

The first independent variable of the content of the mashup as presented to the participant was controlled by the study. Therefore, its ability to potentially influence the participants was assured. Although all participants were told that their mashup contained PGI or PGI + VGI, the degree to which the participants accepted this variable was not controlled. Therefore, in order to help explain the experimental results, participants were asked what information they believed the maps they had just used contained; see Table 6.10.

Table 6.8 Sample size estimations for quality and authority: information as presented

Dependant variable	*Estimated N*	Target ρ	F	ρ
Currency	101	0.00556	(1, 97) = 10.81	0.001
Importance	202	0.02014	(1, 198) = 9.19	0.003
Usefulness	202	0.02014	(1, 198) = 5.79	0.017
Credible	202	0.02014	(1, 198) = 6.45	0.012
Authority	303	0.02014	(1, 299) = 7.05	0.008
Goodness	404	0.02014	(1, 400) = 6.36	0.012
Accuracy	N/A			
Trustworthy	N/A			
Official	N/A			

---

[4] 80 % power, 5 % significance, 2-tailed for Type I and Type III Errors.

6.7 Discussion

Table 6.9 Sample size estimations for quality & authority: information told

Dependant variable	Estimated N	Target ρ	F	p
Authority	202	0.02014	(1, 198) = 9.19	0.003
Credible	303	0.02014	(1, 299) = 7.46	0.007
Usefulness	303	0.02014	(1, 299) = 8.05	0.005
Currency	606	0.02014	(1, 602) = 6.20	0.013
Goodness	606	0.02014	(1, 602) = 6.32	0.012
Official	606	0.02014	(1, 602) = 6.05	0.014
Importance	N/A			
Trustworthy	N/A			
Accuracy	N/A			

Table 6.10 Testing of the 'told' independent variable

Group	Presented	Told	Believed Mashup contained		
			Professional + Volunteer	Professional only	Volunteered only
1	PGI	PGI	18	2	3
2	PGI	PGI + VGI	2	3	28
3	PGI + VGI	PGI	3	14	5
4	PGI + VGI	PGI + VGI	2	1	20

6.7 Discussion

6.7.1 Influence of VGI on Quality and Authority

The first point for discussion is that of the overall effect of presenting VGI alongside PGI within a mashup. Here the presence of VGI in the mashup data was shown to increase judgements of *quality and authority* by a significant amount. This was distinct and separate from the fact that some participants were told that their data included VGI. Based on the results, no support was given to the null hypothesis, and therefore this experiment accepts the alternative hypothesis: "*Presenting groups with maps containing PGI + VGI (rather than just PGI) influences their judgements of quality and authority*". As shown by the comparison of means, this was a positive influence. Consequently, the question that needs to be addressed is whether it was the fact there was *more* information that caused the increase in quality and authority perceptions, or the *unique attributes of VGI* that caused the change in perception.

In a study examining user perceptions of Wikipedia using the information judgement framework of Rieh (2002), Yaari et al. (2011) highlighted how increasing the amount of information available to the user increased perceptions of *quality* and *authority*. This outcome was consistent with the findings of Tillotson (2002), although in a more general study involving university students' assessment of online information. However, in these studies it was the increase in quantity of the same kind of information that caused the increase in perceptions. As shown in the analysis of data used within this study (Chap. 7) while the VGI and PGI data did not conflict with each other, and both focused on the same locations, the issues identified and the way they are described are very different. Therefore, within the context of this study, the VGI data did add additional information, but it was additional information unique to VGI. This suggests that if the same quantity of additional information was provided, yet the information was additional PGI not VGI, then less increase in information judgement would be observed. However, further experimental research is required to fully understand this outcome better.

The second point for discussion was the lack of statistically significant influence of the independent variable *telling people the contents of their mashup*. Therefore, based on the results, this experiment accepts the null hypothesis that there is *"no significant interactions in the quality and authority judgements of maps when users are told they contain PGI or PGI + VGI"*. The question therefore stands as to why telling participants that their mashups contained data from other wheelchair users made no statistically significant interactions in terms of quality and authority judgements.

The first consideration, as demonstrated within Sect. 6.6.2 (Sample size estimation), is that a sample of 202 participants was predicted to be needed in order to produce a statistically significant difference; an achievable sample size. This suggests that the independent variable of telling participants the content of the mashup did have a valid and realistic (but relatively minor) influence on the participants. However, under the experimental design of this study, that difference could not be demonstrated. Further research in this area with larger sample sizes is needed in order to take this investigation further.

The second consideration as to why this independent variable did not achieve statistically significant results is in the limitations of the experimental design. Table 6.10 shows that when asked, many participants did not correctly identify the content of the map as told to them during the tutorial; e.g. told PGI + VGI, believed only PGI. This is despite being told numerous times during the tutorials that their mashup would contain X. Unfortunately, due to the low sample size it was not possible to remove the *noise* in the data set generated by those participants who did not respond to the independent variable of 'being told the content of the mashup' as desired. This noise in the data could be the reason why non-significance was found relating to this independent variable. Unfortunately, as qualitative data relating to *why* the participants believed the content of the mashups to be what it is was not collected, further enquiry on this matter cannot be taken at this stage. This does suggest that simply telling participants the content of the mashups through text and video was not a powerful enough communication method to sufficiently influence their beliefs. However, an alternative proposition could be that participants could not remember what they had been told in the tutorial, or had thought during the map use. It may be speculated that this is consistent with the theory of false memories, where participants recall memories, which are different to the ones held at the time of the event (Gallo 2006).

6.7 Discussion

A third consideration for the non-significance of the Independent variable was that irrespective of the influence of sample size or self-reported beliefs, the influence of the variable was relatively weak. If this is the case then the null hypothesis relating to this variable would be accepted. What is important is that the information itself provides the utility in an easily accessible and understandable fashion.

Consequently, the data appeared to indicate that there should be little concern about utilising VGI (or making users aware that their mashup contains VGI) for fear that it would dissuade the consumer from utilising the map product. As highlighted previously, the additional information within the mashup is most likely causing the increase in quality and authority perceptions. However an important point is that this information can only come from volunteers. Additionally a designer should not look to utilising VGI with the hope of such a *crowd sourced* label increasing user confidence or perceived quality or authority. Instead, the designer must focus on the utility and communication of all potential information sources, selecting the most appropriate one for the user group in a case-by-case situation.

Finally, consideration should be given to the medium effect size associated with the statistically significant outcome that presenting users with VGI alongside PGI increases judgements of *quality and authority*. This score represents the user's overall perception of *quality and authority* as statistically determined within this study. Therefore, it may be inferred that within mashups presenting a mixture of VGI and PGI to a user whose information search is highly dependent on time-sensitive information (e.g. as for kayakers), positive judgements of the website as a whole should increase. This is irrespective of whether they know VGI is included or not. However, this increase may be minor, and not enough for a dramatic change in the way the website is seen and interacted with. Importantly, this was as a result of the increased *quality* of the data in the mashup, and not due to the user *perceiving* the mashup to be better for the simple reason of it includes volunteered data.

6.7.2 Influence of VGI on Currency

Presenting users with VGI alongside traditional PGI (irrespective of what they were told) produced a significant and positive influence on judgements of currency with a medium effect size. Within the investigation into the role that presenting VGI has on user judgements, this was the most influential component of the experiment. This outcome is contextualised by Goodchild (2008), who commented that *"perhaps the most significant area of geospatial data qualities for VGI is currency, or the degree to which the database is up-to-date"*. Consequently, the question needs to be asked, 'why was currency influenced by the VGI data?' While the VGI collected during the data generation chapter was undoubtedly *more current* than the PGI collected through literature review with regards to intermediate or fast changing information, comparison of the data sets showed no demonstrable disagreement.

Currency in this sense relates to the objective currency of information; e.g., the information can be demonstrated to reflect the current state of the environment and therefore utility may be derived. Under this definition, PGI may be seen as current, although this is particularly true when relating to static information. Currency has also been highlighted as an important dimension of a user's perception of online information (Flanagin and Metzger 2007; Metzger et al. 2003). Within the current literature, a significant body of work (Barry and Schamber 1998; Goodchild 2007)—as well as Study Two within this book—demonstrated the important connection between the *currency* of information and VGI. This finding may be explained by the work of various authors (Gitelson and Crompton 1983; Nolan 1976; Schuett 1993) who demonstrated that information from *informal sources* is the most informative due to its ability to reflect changes in the environment. This benefit is, however, limited to where the data describes events and geography that changes faster than traditional cartography can document, or relate to information not captured by traditional PGI.

As the discussion within the data generation chapter demonstrated, the VGI collected (and presented to users within this study) contained not only objective data which could be achieved through traditional/professional methods, but also experiential and emotional data which can *only come* from users. PGI, however, covered more objective features; e.g. station is step free for easy wheelchair access. This means the VGI and the context of use within this study can be considered *informal* in the way which Gitelson and Crompton (1983), Nolan (1976) and Schuett (1993), meant it. This may explain why the participants in this study judged VGI enhances mashups to be of higher currency. However, further research to confirm this application of theory is required.

Finally, the quantitative approach to research within this study did not provide sufficient data to infer why presenting VGI alongside PGI to participants only seemed to influence the judgements related to currency. It would have been useful to have obtained more qualitative data on: the perceptions of participants towards the different versions of the mashups, the extent to which participants were fully aware of the presences or otherwise of VGI, and the benefits (if any) that they thought it conveyed.

6.7.3 Sample Size Estimation

While the sample size within this study passed the minimum assumptions required by MANOVA, the lack of significant outcomes may have been due to the number of participants in each case being too small to detect relatively small differences due to the manipulation of the independent variables. Therefore, sample size estimation was utilised in order to predict potential outcomes with increased sample sizes.

The most useful and robust outcome from this processes was that should the experiment be re-run in the future, a minimum of 400 (and ideally 600)

6.7 Discussion

participants should be sought. This would allow for the full range of influences on the dependable variables coming from the independent variables to be assessed. This supports the claims of Borg and Gall (1989) that 100 participants per cell are required for robust statistical analysis. Further to this, sample size analysis has allowed for a number of inferences to be drawn that describe what *may* be found if sufficient participants were to be sourced.

6.8 Conclusions

Through investigation, this study has addressed the study aims in the following ways:

1. *The extent to which including VGI within the mashup alongside PGI affects the users' judgements*
   Although VGI has a great potential to contain and represent a wide range of data not easily captured by traditional techniques (Burns 2009; Goodchild 2007; Kingsbury and Jones III 2009), its influence on user judgements within a simple, online mashup was limited. Within this context, including VGI within a data set has been shown to increase *quality* and *accuracy* judgements by a statistically significant amount. Here, the independent variable of perceived currency is the most sensitive the inclusion of VGI. Consequently including VGI alongside PGI in a mashup may enhance the user experience by a small yet noticeable amount, and without any negative impact on user perceptions.
2. *The extent to which knowing that mashups contain VGI influences user judgement*
   Telling users that their mashup contained VGI through embedded video and on-screen text had no statistically significant influence on the judgements of users. This suggests that the supposed assumption that users feel VGI is inferior to PGI (and thus knowledge of its use is detrimental to the user experience) may not be true.
3. *The extent to which understanding user reactions to VGI and PGI may influence the design of future mashups*
   While the inclusion of VGI within the data set may not necessarily produce a large benefit in the way users perceive the website, this study has highlighted the subtle yet beneficial ways in which VGI may enhance the user experience. Additionally, this study demonstrates how VGI is limited in what it may be able to achieve, so informing the designer to search for other, more efficient and useful ways at enhancing those elements that VGI is not able to enhance by a noticeable or useful amount. This study has shown that while the presentation and promotion of VGI are important and useful for the design of high quality user experiences in neogeography, consideration is needed as to the area of user judgement that may be enhanced. A human factors designer should consider if

the potential gains of utilising VGI are *worth* the extra challenges that their successful implementation would bring.

This study supports the combined use of VGI and PGI over presenting just VGI or PGI. However, this study has also highlighted key limitations in the ability for VGI to enhance all areas of *quality* and *authority* within the mashup. Importantly, there were no negative repercussions for the inclusion and utilisation of VGI. This is possibly one of the most interesting outcomes as it answers the concern in the literature on the potential *dangers* of presenting users with information from untrained amateurs.

References

Albaum G (1997) The likert scale revisited: an alternate version. J Market Res Soc 32(2):331–348
Alreck PL, Settle RB (1995) The survey research handbook. McGraw-Hill, New York
Barry CL, Schamber L (1998) Users' criteria for relevance evaluation: a cross-situational comparison. Inf Process Manage 34(2/3):219–236
Bartlett MS (1954) A note on the multiplying factors for various Chi square approximations. J Roy Stat Soc 16(Series B):296–298
Bishr M, Mantelas L (2008) A trust and reputation model for filtering and classifying knowledge about urban growth. GeoJournal 72(3–4):229–237
Boin AT, Hunter GJ (2006) Do spatial data consumers really understand data quality information. In: Caetano M, Painho M (eds) 7th international symposium on spatial accuracy assessment in natural resources and environmental sciences. Lisbon, Portugal, p 215
Borg WR, Gall MD (1989) Educational research. Longman, White Plains
Burns RL (2009) Spatializing places, people, and utterances: a case study involving san diego neighborhoods. San Diego State University, San Diego
Carter S (2011) Statistics show disabled use internet less than nondisabled 2011 (Oct 6th). Available at: http://www.examiner.com/disability-in-dallas/statistics-show-disabled-use-internet-less-than-nondisabled
Caswell F (1995) Statistics. John Murray, Guildford
Catell RB (1966) The scree test for number of factors. Multivar Behav Res 1(5):245–276
Cohen J (1988) Statistical power analysis for the behavioral sciences, 2nd edn. Lawrence Erlbaum Associates, Hillsdale
Collins J (2006) An investigation of web-page credibility. J Comput Sci Coll 21(4):16–21
Comber AJ, Fisher PF, Wadsworth RA (2007) User-focused metadata for spatial data, geographical information and data quality assessments. In: 10th AGILE international conference on geographic information science. Aalborg University, Denmark, pp. 8–11
Devillers R et al (2002) Spatial data quality: from metadata to quality indicators and contextual end-user manual. In: OEEPE/ISPRS joint workshop on spatial data quality management. Istanbul, Turkey: OEEP, p 45
Edsall R (2007) Cultural factors in digital cartographic design: implications for communication to diverse users. Cartography Geogr Inf Sci 34(2):121–128
Ericsson KA, Simon H (1993) Protocol analysis: verbal reports as data. MIT Press, Cambridge
Field A (2004) Discovering statistics using SPSS for windows. Sage Publications, Thousand Oaks
Flanagin AJ, Metzger MJ (2007) The role of site features, user attributes, and information verification behaviors on the perceived credibility of web-based information. New Media Soc 9(2):329–342. Available at: http://nms.sagepub.com/content/9/2/319.abstract

References

Fox S (2010) The promise of mobile as part of the "Managing Health Care Information" track. In: Transform 2010. Minnesota, USA: Centre for Innovation. Available at: http://centerforinnovation.mayo.edu/transform/2010

Frank AU (1998) Metamodels for data quality descriptions. In: Jeansoulin R, Goodchild MF (ed) Data quality in geographic information: from error to uncertainty. Paris, France: Hermès, pp 15–29

Frick A, Bächtiger MT, Reips U-D (2001) Financial incentives, personal information and drop-out rate in online studies. In: Reips UD, Bosnjak M (eds) Dimensions of internet science. Pabst Science, Lengerich, pp 209–220

Gallo DA (2006) Associative illusions of memory: false memory research in DRM and related tasks. Psychology Press, New York

Gitelson RJ, Crompton JL (1983) The planning horizons and sources of information used by pleasure vacationers. J Travel Res 21(3):2–7. Available at: http://jtr.sagepub.com/content/21/3/2.short

Goodchild MF (2007) Citizens as sensors: the world of volunteered geography. GeoJournal 69(4):211–221. Available at: http://www.springerlink.com/content/h013jk125081j628/

Goodchild MF (2008) Spatial accuracy 2.0. In: Zhang J-X, Goodchild MF (eds) Proceeding of the 8th international symposium on spatial accuracy assessment in natural resources and environmental sciences. World Academic Union, Shanghai, pp 1–7

Idris NH, Jackson MJ Abrahart RJ (2011b) Map mash-ups: what looks good must be good? In: Emma Jones C et al (eds) Proceedings of the 19th GIS research UK annual conference. GIS Research UK, Portsmouth, UK, p 119

Idris NH, Jackson MJ, Abrahart RJ (2011a) Colour coded traffic light labeling: a visual quality indicator to communicate credibility in map mash-up applications. In: Presented at the international conference on humanities, social sciences, science and technology (ICHSST), Manchester, UK, pp 1–7. Available at: http://www.geoinfo.utm.my/geoinformatic/geoinformatic publications/2011/Nurul_Idris_ICHSST_new.pdf

Kaiser HF (1970) A second generation Little Jiffy. Psychometrika 35(4):401–415

Kaiser HF (1974) An index of factorial simplicity. Psychometrika 39(1):31–36

Keller DK (2006) The Tao of statistics. Sage Publications, Thousand Oaks

Kingsbury P, Jones JP III (2009) Walter Benjamin's Dionysian adventures on Google Earth. Geoforum 40(4):502–513

Levine BS et al (1993) A national survey of attitudes and practices of primary-care physicians relating to nutrition: strategies for enhancing the use of clinical nutrition in medical practice. Am J Clin Nutr 57(2):115–119

Maguire M (1998) SAQ—software acceptance questionnaire—developed as part of the telematics application (TA) CEC Respect project, process of quantitative assessment of user acceptance edn. HUSAT Research Institute, Now part of Loughborough Design School, Loughborough University, UK

Metzger MJ et al (2003) Ringing the concept of credibility into the 21st century: integrating perspectives on source, message and media credibility in the contemporary media environment. In: Kalbfleisch PJ (ed) Communication yearbook 27. Taylor and Francis, Mahwah, pp 293–335

Mizumoto A, Takeuchi O (2009) Comparing frequency and trueness scale descriptors in a Likert scale questionnaire on language learning strategies. JLTA J 12:116–136

Mummidi L, Krumm J (2008) Discovering points of interest from users' map annotations. GeoJournal 72:215–227

Murphy KR, Myors B (2004) Appendix D. In: Statistical power analysis. Lawrence Erlbaum Associates, Mahwah

Nisbett RE, Miyamoto Y (2005) The influence of culture: holistic versus analytic perception. TRENDS Cogn Sci 9(10):467–473

Nolan SDJ (1976) Tourists' use and evaluation of travel information sources: summary and conclusions. J Travel Res 14(3):6. Available at: http://jtr.sagepub.com/content/14/3/6.short

O'Reilly T (2005) What is Web 2.0: design patterns and business models for the next generation of software. oreilly.com, 2010 (Dec 27th), pp 1–5. Available at: http://oreilly.com/web2/archive/what-is-web-20.html. Accessed 2 April 2013

ONS (2010) Internet Access. Homepage of the office for National Statistics, 2011 (June 22nd). Available at: http://www.statistics.gov.uk/cci/nugget.asp?id=8. Accessed 22 June 2011

OUP (1989) The Oxford English dictionary. Oxford University Press, Oxford

Pallant J (2010) SPSS survival manual fourth. McGraw-Hill Education, Maidenhead

Parker CJ (2011) Map Pro 1: first map in pro series. UMAPPER. Available at: http://www.umapper.com/maps/view/id/102633/. Accessed 22 July 2013

Parker CJ, May AJ, Mitchell V (2012) Using VGI to enhance user judgements of quality and authority. In: Whyatt D, Rowlingson B (eds) Proceedings of GIS research UK 20th annual conference. GIS Research UK, Lancaster, UK, pp 171–178

Reips UD (1996) Experimenting in the world wide web. In: Proceedings of the 26th society for computers in psychology conference (SCiP–96). Chicago, USA: SCiP

Reips U-D (1999) Theorie und Techniken des Web-Experimentierens [Theory and techniques of web experimenting]. In: Batinic B et al (eds) Online research: Methoden, Anwendungen und Ergebnisse. Hogrefe, Göttingen

Rieh SY (2002) Judgment of information quality and cognitive authority in the Web. J Am Soc Inform Sci Technol 53(2):145–161. Available at: http://onlinelibrary.wiley.com/doi/10.1002/asi.10017/full

Rieh SY, Belkin NJ (2000) Interaction on the web: scholars' judgment of information quality and cognitive authority. In: Proceedings of the 63rd ASIS annual meeting, vol 37, pp 25–38. Available at: http://rieh.people.si.umich.edu/~rieh/papers/rieh_asis2000.pdf

Schuett MA (1993) Information sources and risk recreation: the case of whitewater kayakers. J Park Recreation Adm 11(1):67–77. Available at: http://js.sagamorepub.com/jpra/article/view/1796

Shi Q (2010) An empirical study of thinking aloud usability testing from a cultural perspective. An empirical study of thinking aloud usability testing from a cultural perspective. Doctoral Thesis

Tabachnick BG, Fidell LS (2001) Multiple regression. In: Pascal R, Brown W (eds) Using multivariate statistics, 4th edn. Allyn and Bacon, Needham Heights

Tillotson J (2002) Web site evaluation: a survey of undergraduates. Online Inf Rev 26(6):392–403. Available at: http://www.emeraldinsight.com/journals.htm?articleid=862200&show=abstract

Watkins MW (2000) Monte Carlo PCA for parallel analysis. Available at: http://www.softpedia.com/get/Others/Home-Education/Monte-Carlo-PCA-for-Parallel-Analysis.shtml

Weathington BL, Cunningham CJL, Pittenger DJ (2010) Research methods for the behavioral and social sciences. Wiley, USA

Yaari E, Baruchson-Arbib S, Bar-Ilan J (2011) Information quality assessment of community generated content: a user study of Wikipedia. J Inf Sci 37(5):487–498

Chapter 7
Conclusion

7.1 The Nature of VGI as Distinct from PGI

7.1.1 Data Content, Use and Contribution

Various insights into the differences between VGI and PGI in terms of their content have been observed. Studies One and Two highlighted how VGI and PGI may vary in their use of standardised *terminology, frequency of surveying/resurveying areas* and *quality control* (amongst others). However, from the perspective of the consumer-user, a distinction is not made between VGI and PGI, and they are instead seen as simply information. Here, the *volunteer* or *professional* originator has little impact on the consumer's use and assessment of the information. While such a point was originally contested (Das and Kraak 2011; Metzger and Flanagin 2011; Flanagin and Metzger 2008), study Three demonstrated how informing participants that their mashups contained data from amateur volunteers had a largely negligible (although positive) impact on user judgements. What was shown to have the greatest positive influence on user judgements was including VGI within the data set, irrespective of whether the user believed the mashups data to have been generated by professionals or volunteers. It is unlikely that this was the result of VGI simply offering more information, since the judgement dimension which was the greatest influenced by the inclusion of VGI alongside PGI was that of currency.

While providing more information may influence the judgements of credibility, trustworthiness or authority, currency judgements are not based on quantity. Instead, they are based on the ability to reflect current events, demonstrated by the way the information is written and presented (Alonso et al. 2007; Schilder and Habel 2001). Therefore, consideration must be given to the characteristics of the information which influences the user's judgement, more so than the level of professionalism accredited to the contributor.

Due to its standardised and well-documented approach, the creation of PGI is a well-understood field, with processes catalogued and discussed in detailed within the literature (Crone 1968; Monmonier 2006; Ordnance Survey 2009). However, the production of VGI is a more elusive and less understood subject, possibly due to its anarchic nature and it being a relatively recent phenomenon. Additionally, each VGI project takes a unique approach on crowd sourcing for its information data set, and therefore the search for a universal description of such a process is elusive. Goodchild (2007a) propositioned the use of the world's six billion inhabitants as potential contributors of VGI, the implication being that anyone may be and could be a VGI contributor. However, as study Two demonstrated, there currently exists a large gap between the use of VGI and the desire to generate VGI. In fact, many of those whose activities relied heavily on the anonymous contributions of others did not see the act of sharing their own experience as important. Within the framework of the scoping study, such attitudes were clearly different between those who belonged to *Special Interest Mapping Groups* (and were keen to contribute and develop GI) and those who were *consumers* of GI and had no interest in the development of the source. In general, the pervasion of smartphone and crowdsourcing in society has been gathering momentum since the mid 1900's (Alonso et al. 2008; Doan et al. 2011; Tapscott and Williams 2008). However, study Two demonstrated that the information originating from volunteers which has the greatest impact on the outcomes of user activities is that which may offer personal perspectives and opinions. This can contextualize complimentary information, which may be PGI. Therefore, the act of producing and contributing widely effective and useful VGI is the role of the purposeful individual who strives to do so for a possibly unspecified reason.

This book has highlighted how although volunteers can come from any location and background, it takes a certain motivation or desire in order to drive an individual to contribute. This is in line with the work of Rogers (2003), who showed how simply having knowledge and access to technology was not sufficient indicators for its adoption. Since VGI is a form of crowdsourcing (Goodchild and Glennon 2010; Zook et al. 2010), consideration is needed as to (1) recruitment, (2) user motivation for engagement in the contribution process, (3) long term retainment of contributors and (4) the forms of tasks given to the contributors (Doan et al. 2011; Reeves and Sherwood 2010). While the reasons why members of *Special Interest Groups* are engaged in VGI creation may be explained through these four perspectives, interesting issues arise from the consideration of consumer-users. The scoping study addressed this aspect, highlighting how at the core of their activities, consumers' desire to achieve their goals, with no clear preference or consideration given to whether the data comes from volunteers or professionals. However, as highlighted through study Two, the demonstrated desire to achieve interaction with appropriate data for their own needs presents a opportunities for collection of VGI by consumers. This is a position proposed by Goodchild (2007b). The fact that *consumers* were shown to be more reluctant towards data contribution than *Special Interest Groups*[1] means that although their

---

[1] **Special Interest Group**: Individuals who come together to collaboratively achieve some shared goal (Coote and Rackham 2008).

7.1 The Nature of VGI as Distinct from PGI

engagement in VGI creation is potentially lucrative (Goodchild 2007b), such an effort may be limited. This may be due to the relative complexity of contributing data, the lack of communication from VGI to consumers that contributions are needed, or the lack of motivation for consumers to contribute to projects.

A final consideration relevant to this discussion is the role of mobile computing (e.g. smartphones, tablets) in VGI. Since the start of this book, the smartphone (and the ubiquity of third party apps to take advantage of the hardware and user interface) increased exponentially in pervasion (Alonso et al. 2008; Doan et al. 2011; Tapscott and Williams 2008). Consequently, contribution of VGI has shifted from being a very technical, hands on event, requiring dedicated GPS devices and ability to upload their traces (see Chap. 3) to a simple, and interactive event using third party apps. This has allowed the number of contributions to increase, and more specialist groups such as wheelchair users to volunteer their information; see Fig. 7.1. Additionally, comparable websites such as AccessAdvisr (www.accessadvisr.net) have started collecting subjective VGI (e.g. how friendly were railway station staff?) alongside subjective matters (e.g. does the railway station have stepped entrances).

Consider this, it may be expected that future crowd sourced and VGI projects will fully embrace the subjective information which can only come from persons to whom the information relates to. Since this book has demonstrated that (within the contextual limitations of the studies) it is the qualitative and subjective information which is the greatest strength of VGI over PGI, then these projects look to become more prevalent, influential and important within society over the next few decades. However, such future gazing is outside the scope of this project.

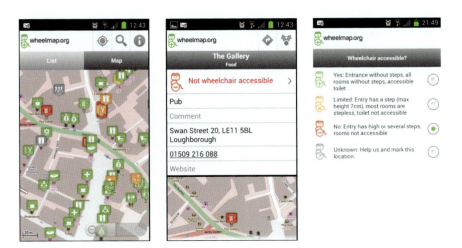

Fig. 7.1 Screen shots of the android wheelmap app (SozialHelden 2012)

7.1.2 Information Judgements

The scoping study demonstrated that classifying users by their use of geographic information was an effective way to group individuals and organisations in order to understand their attitudes towards VGI. This stemmed from the categorisation system of Coote and Rackham (2008) that VGI users exist as *Consumers, Special Interest Groups (SIG), Local Communities (LC)* or *Professionals*. It was shown that the user's use of mashups and information, requirements on *accuracy, relationship to other users,* and *personal ideological bias* can create unique perceptions towards VGI on a group-by-group basis. Therefore (within the limitations of the studies within this book) if a neogeographic product was produced to fit the user attitudes, requirements and interaction preferences of a *SIG*, it may be unsuitable for consumers, LCs or professionals. For example, the OpenStreetMap mapping platform JOSM has been at the centre of data contribution to the project from very early on, yet for all its advances over the years, remains largely inaccessible in terms of usability to anyone without the time or dedication to learn to use it. This may cause the product to be rejected outright as unfit for purpose by such user groups.

This philosophy of developing the product with the users, their characteristics and their needs is one of the most fundamental approaches and themes within Human Factors and Interaction Design (Burns and Vicente 1996; Flach et al. 1998; Norman 2005; Preece et al. 2002). Although the literature relating to VGI provides a relatively useful perspective on user needs (Goodchild 2007a; Obermeyer 2007), the users in the literature tend to be treated as a homogenous block (e.g. 'the users') rather than as separate and distinct groups (e.g. the consumer-user). Therefore a more detailed understanding of user perceptions related to VGI would allow for current and past work to be contextualised so that user centred design practice can be applied. A limitation to this book is how investigations following the scoping study investigated only the position of the consumer-user. An interesting outcome was how study Two (Sect. 4.5.1.6—*Relevance of Information Sources*) hypothesised that the more knowledgeable and *accurate* an information source is (in the sense of reflecting the conditions of reality in line with how the information searcher will experience them), the more likely it is to be seen as authoritative and professional. It was also suggested that in this situation it is *accuracy* rather than a *logo* that may be emphasising professionalism. This is interesting since in Study Three, adding VGI to the mashup data (and presumably being assessed as knowledgeable and accurate data) did not increase perceptions of authority or professionalism. However, these two studies focused on different tasks and use situations, to which the generalisability is currently unproven by complimentary studies. This difference between the two studies may be because the VGI included within the mashup did not contain the right attributes to be considered *more credible* than the PGI, when considered from the viewpoint of the user (Wilson 1983). Alternatively, it may be that the VGI did not add sufficient increases in *usefulness, goodness, currency* or *accuracy* (Rieh 2002) to cause an

7.1 The Nature of VGI as Distinct from PGI

effect. It is clear that further research is required in order to better understand whether adding VGI to a data set increases its perceived authority based on the user having knowledge of the contributor(s) of the data. If such an experiment was to be conducted, a central theme must involve the self-selection of information. This is because it is possible that the participants in Study Two perceived VGI as being authoritative since they *chose* the VGI sources they talked about, whereas participants in Study Three had no choice over the information they had to consider.

7.2 An Appraisal of the Framework of VGI

While Sect. 7.1 (above) highlighted the differences between VGI and PGI from a human factors perspective, this section aims to discuss the Framework of VGI proposed within Chap. 4 (see Fig. 7.2). This can provide a framework with which to further understand the scope of VGI and its role in products.

The framework as presented in Fig. 7.2 proposes that the two most important factors when making the distinction between VGI and PGI projects is the *objectivity* and *quality control* as demonstrated in the data; discussed below.

The framework presented above shows both VGI and PGI together within a single framework—as opposed to producing two complimentary frameworks for VGI and PGI respectively. Doing so is in line with the findings within this book

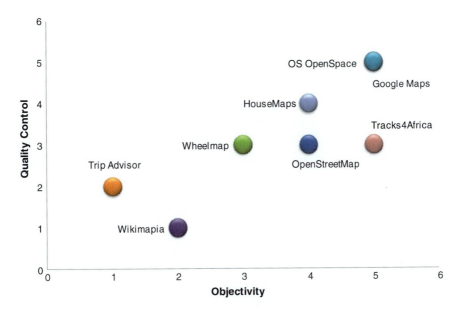

Fig. 7.2 A proposed framework for neogeographic products

that demonstrated how *consumers* utilise VGI and PGI alongside each other with the same critical requirements. In particular, while telling users that their mashups contained VGI had almost negligible impact, improving the mashup data set with VGI generated noticeable and interesting influences on user judgements within Study Three. This is contrary to the standard practice within the literature, where VGI and PGI data sets are often referred to as two different and largely incompatible forms of information (Flanagin and Metzger 2007; Keen 2007; Tsou 2005; van Exel and Dias 2011). However, Study Two demonstrated how consumer-users utilise both VGI and PGI sources alongside each other in order to manage the level of risk in their activity. Additionally the degree of trust they placed in information was greater when coming from multiple sources being used together to converge on a *truth*. Importantly, the criteria used to assess their discovered information were the same for both VGI and PGI sources. This is complimentary to the outcome from the scoping Study—that consumer-users require their information to aid them in their activities, irrespective of its volunteer or professional origins. However, it must be noted that the scoping Study also demonstrated how each user group perceives VGI and PGI differently. Here, the higher their personal investment in an information source the more biased they are towards it; e.g., an OpenStreetMap contributor is biased towards OpenStreetMap. Therefore, while presenting VGI and PGI alongside each other within the framework can be considered sound from the perspective of the consumer-user, it may have limited applications in describing the way in which professionals, *Special Interest Groups* or local communities relate to VGI and PGI.

Objectivity and *quality control* are shown as the two-categorisation elements of the matrix. Within this book a broad definition of quality has been taken as *the extent to which the product or service satisfies the technical or specific needs of an individual or organisation*. This is different from the notion of *Quality Control*, being the processes of examining a product or system to determine whether or not it accomplishes was what was specified by the designer in the design (DeGarmo et al. 2003). However, quality control has been shown to influence the overall quality of GI, and is therefore a useful predictor of the user's perception of the information's *quality* (Bevan 1999; DeGarmo et al. 2003). Viewing the rich picture of the scoping Study (Fig. 3.2, p. 43) the key concerns of the users, while unique to each group, may be categorised as (1) concerns about the content of the maps relative to user needs and (2) concerns about trust, or the degree to which the information is correct. Under the definition as used in the book, these may be drawn together through the consideration of the information's *quality*. Information of relatively high *quality* may be assumed by the user to have fewer issues associated with *accuracy* than that of low quality. Finally, in the literature quality has been discussed as a key issue which has yet to be mastered relative to VGI, but once done so shall provide a useful and effective categoriser (Goodchild 2008; Bishr and Mantelas 2008; Mummidi and Krumm 2008).

At this point it is worth exploring the relation between quality control and findings of this book. As shown in studies Two and Three, the inclusion of volunteered information in a mashup does not lower the quality or authority of neogeography, as was the concern of several authors (Flanagin and Metzger 2007; Keen 2007; Tsou 2005; van Exel and Dias 2011). Additionally, Haklay et al. (2010a) and Holone et al. (2007) both demonstrated that the more an instance of VGI is edited, the higher quality it becomes. This leads onto a question of how quality control may be introduced into a system that Budhathoki et al. (2008) described as anarchic. The first point is that saturation of results in the data generation chapter supports the proposition of Haklay et al. (2010a) that five participants editing an instance is sufficient to produce a quality data set. Secondly, Bishr and Mantelas (2008) showed that VGI comes in a degree of *qualities* and should be filtered to ensure high quality content is presented. Therefore, a simple metric could be constructed for a mashup where the map may be edited by any individual, yet the instance which is being edited would not be available to consumers until a minimum of five edits has been encountered. This may allow for group consensus to emerge through the anarchic scene of VGI, forming an effective although untraditional method of quality control. However, the drawback to such an approach would be counter to the principals of crowd sourcing engagement, which stresses instant visible feedback to the contributor as a *reward* (Mihalcea and Chklovski 2003).

Objectivity is the second categorisation term within the framework. When users search for information that describes an area of interest to them in terms of good, bad, difficult (etc.), then subjective information is of most use. While not experimental or subjected to rigorous testing, the data generation chapter demonstrated various differences between VGI and PGI, with their levels of objectivity being an important and central outcome. This observation is supported by Study Two in how users sought a combination of subjective and objective information in order to converge on a *truth* about the environment relative to their needs. Additionally, Study Three showed how adding VGI alongside PGI increased with perceptions of quality and authority, most likely as a result of the inclusion of subjective opinions (VGI) alongside objective statements (PGI).

It is therefore the conclusion of this section that the framework as presented within this book provides a potentially useful way to discuss neogeographic projects in relation to one another. However, while the suitability of its two dimensions are supported studies into information use (Study Two) and judgement (Study Three), it has not yet been utilised, tested and developed within a design context.

7.3 Unique Influences of VGI on the User

This book has shown how both VGI and PGI play particular roles within online information search. In particular, while PGI sources may effectively describe relatively static objects (e.g. trees, building locations, topography, etc.), VGI comes from a convergence of amateur sources, with each source describing specific points that are perceived by the author to be of interest to others. Additionally,

VGI was shown to cover a wider range of topics than PGI, although it was of most use when describing niche subjects in detail far greater than achieved in PGI. However, it is important to note how this is limited in applicability to the tasks and contexts of the studies of this book.

An obvious importance of this is VGI being able to capture and produce data sets not possible under traditional cartographic means—as was highlighted by the *Special Interest Groups* within the scoping Study. The impact of this convergence of multiple sources on a wider reaching *truth* as described in Study Two was measured and understood within Study Three. While the benefits as measured were not as profound as some of the current literature may have assumed (Grira et al. 2010; Ray and Ryder 2003; Tapscott and Williams 2008), including VGI alongside PGI had a definite and positive influence on the overall perception of the mashups *quality and authority* from the position of the user; particularly *currency*. While this is indicatory of a wider trend, such an outcome should only be applied with confidence to online mashups delivering transport accessibility information.

Although further research is required within experimental settings and different use contexts, Study Three demonstrated that VGI can positively influence the information judgements of users. Further to this, the reason why VGI influences the judgements of users is its difference to PGI. Here, the PGI fills a need where this traditional form of information excels into a lower degree; as shown within Study Two. Moreover, the influences of VGI as described within Study Three (above) are highly compatible with the concerns and tensions of consumer users presented within the scoping Study. For example, the consumer concern for trust in the data provided to them may be addressed by the increased *currency, credibility* and *usefulness* of the data as derived from VGI. Therefore, this book has been able to describe the benefits of VGI (Study Two), the ways they influence user judgement (Study Three) and how they address the concerns and needs of consumer users (The Scoping Study).

7.4 Limitations of VGI from a Human Factors Perspective

Human factors was broadly defined by Burns and Vicente (1996) as being concerned with the design of artefacts to be consistent with a human user's physical and psychological capabilities. More importantly, Norman (2005, p. 124) wrote that *good behavioural design should be human-centred, focusing upon understanding and satisfying the needs of the people who actually use the product*. To take a human factors perspective is therefore to design what is best for the user in terms of their technological, personal (user), control or use requirements (Flach et al. 1998). While each of these design views offers different perspectives on the user-product relationship (helping the designer to produce highly functional products to their project's specification), this book relies upon the framework of user centred design to generate research results that are relevant to future products/ services incorporating VGI. Taking this angle directs focus away from the

7.4 Limitations of VGI from a Human Factors Perspective

technological and physical limitations of VGI and its production, and towards to the relationship between the user and the amateur volunteered information.

Goodchild (2007a) commented that VGI is able to provide information at a faster rate than traditional methods, filling a long-standing gap in cartography (Crone 1968; Wood 2003). Such a proposition is supported by this book since the scoping Study demonstrated the acceptance of VGI by users from consumer to professional, Study Two highlighted VGI's great strength in providing current information, and Study Three showed how VGI may enhance judgements of mashups being current, and of high quality and authority. Present literature has also highlighted that VGI may play an important role in fulfilling the call for more specialist maps (Goodchild 2007b; Crone 1968), or achieving a diversity in GIS previously not possible due to commercial viability (Goodchild 2008; Pultar et al. 2008). However, this book has been able to build upon much of the speculation and suggestion of previous research, demonstrating potential as well as placing limitations on the ability of VGI to be an information *addition* to neogeographic systems to enhance the user experience. The above may be given a deeper perspective by the outcome from the scoping Study that *consumers* (those who primarily utilise) and *Special Interest Groups* (those who primarily contribute) are fundamentally different in their attitudes and relationship towards VGI products. Consequently, while almost anyone *can* contribute data (Goodchild 2007b; Shirky 2009), only a self-nominated minority *will*. This is mirrored within Study Two of this book, where in the context of outdoor recreation, people were more willing to view and receive information than actively share and disseminate their experiences to help others. However, the degree to which this impacts upon the utility of VGI is debatable, since small numbers of contributors may make large and effective data sets (Bishr and Kuhn 2007; Haklay et al.2010a).

From a theoretical point of view, Petty and Cacioppo (1986) and Idris et al. (2011a) identified that people are not always motivated to scrutinise every message that they come across. Additionally Warnick (2004) demonstrated that over time the source of the information is decreasingly of use within determining information credibility. This may explain why presenting users with VGI had a greater impact on their information judgements than telling them that their mashups contained VGI. The benefits to the user are closer associated with the functionality of the data relative to the user needs, rather than the perceived image of the data author. This in turn relates to the concept of information *value* as being derived relative to the needs of the user (Badenoch et al. 1994), how it reduces uncertainty (Sheridan 1995) and its ability to make a difference (Bateson 1988; Koops 2004; Stephens 1989). From this it may be seen how if utilised in the correct fashion by a designer, VGI alongside PGI may increase perceived value of a neogeographic product while producing greater usability; as defined by ISO 9241-11 (1998).

VGI is of most use when it describes the world in ways that PGI cannot. However, this is relative to the needs of the user, rather than a demonstrable geographic or cartographic specification. This therefore raises the question of the reusability of VGI outside the context it was created for. Unfortunately, this book does not explicitly tackle this issue. However, the wide variety of formal and informal sources described within Study Two suggests that while the further away from the intended contribution the VGI is used the less effective it is, VGI may be able to be effectively utilised within a number of contexts. A popular example of this is volunteer mapping projects such as Google Map Maker or OpenStreetMap producing the best maps for less economically developed countries where national mapping agencies are ill equipped to tackle the substantial cartographic challenge (Cooper et al. 2011; Zook et al. 2010). However, these map generation forms of VGI limit the user tasks to the contribution of largely objective information, and therefore limit the scope of their related perspectives to the wider field of VGI. Additionally, while Study Two highlighted VGI's limited ability to describe large geographic areas to the precision and coverage to which PGI has traditionally excelled at. However, their suitability to this scale depends greatly on task which the user is searching information for in order to achieve. As value is derived from the use of data in specific contexts (Badenoch et al. 1994), the generation of theory to describe or predict such potential may be elusive. However an approach of such a theory could be that the more niche the object that the information describes, and the faster it alters its conditions, the less transferable that information is. While this may be a limitation in VGI, it also presents an opportunity since these are the conditions identified as opportunities for VGI to provide benefit to a specific user group.

7.5 Design Recommendations for Utilising VGI

Due to the limitations of this book as derived from the restricted number of user tasks considered through the research chapters, design guidelines in this section should be taken as indicatory rather than mandatory. As highlighted within the introduction of this book, at the time of submission there is a lack of guidelines on how to develop and evaluate mashups (Idris et al. 2011a).

- Cover the widest range of consumer-user information needs by combining VGI and PGI alongside each other in a neogeographic system or mashup.
- While judgements of *quality, authority and credibility* have been shown to be positively influenced by the inclusion of VGI within a mashup, neogeographic designers may need to find alternative ways of influencing the user's holistic judgements of the online information, since the simply including VGI within a mashup will not alone create the *killer app*.
- To use information most appropriately, use PGI to describe general, permanent and objective features of the landscape (e.g. location of a castle), and VGI to describe specific features in depth related to the subjective opinions of the associated user group (e.g. 'easy access' to all areas of the castle for wheelchair users).

7.5 Design Recommendations for Utilising VGI

- To capture highly relevant experience from users and thus improve the VGI data set, seek to promote contribution of experiences and opinions as a natural and purposeful part of the neogeographic system.
- Take into account the activity within the VGI contributor community, and ensure that it is lively enough for erroneous or incorrect data to be corrected or updated by fellow contributors.
- Allow a clear and easily accessible comparison between multiple information sources (VGI and PGI) within the mashup to allow users to converge on a common truth and find the mashup more useful, effective and satisfying. For example, if a developer was producing a mashup of accessibility information, a degree of benefit to the user would be found by collating professional information sources. However, by adding to this the voluntary contributions of amateurs (e.g. parents with prams, wheelchair users, etc.) then the mashup would cover a wider spectrum of issues a user may face while navigating the built environment.
- Favour the use and utilisation of VGI and PGI information sources that take advantage of the pervasive data capture and representation inherent in Web 2.0 technologies. For example, the name of streets, places and shops form an important dimension to geographic information and provide a useful context within mashups and information delivery. However, rather than being static, they alter and change at a rate faster than traditional techniques can accommodate (Monmonier 2006). Instead of treating such information as if it was a static geographic feature (such as a road) and instead allow Web 2.0 technologies to constantly update this frequently changing data would provide a wealth of additional accuracy and context to a mashup.

References

Alonso O, Gertz M, Baeza-Yates R (2007) On the value of temporal information in information retrieval. ACM SIGIR Forum 41(2):35–41. Available at: http://dl.acm.org/citation.cfm?id=1328968

Alonso O, Rose DE, Stewart B (2008) Crowdsourcing for relevance evaluation. ACM SIGIR Forum 42(2):9–15

Badenoch D et al (1994) The Value of Information. In: Feeney M, Grieves M (eds) The value and impact of information. Bowker-Saur Limited, Chippenham, pp 9–78

Bateson G (1988) Glossary. In: Mind and nature: a necessary unity. Bantam Books, New York, p 245–249

Bevan N (1999) Quality in use: meeting user needs for quality. J Syst Softw 49:69–89

Bishr M, Kuhn W (2007) Geospatial information bottom-up: a matter of trust and semantics. In: Irina-Fabrikant S, Wachowicz M (eds) The European Information Society. Springer, Berlin, pp 365–387

Bishr M, Mantelas L (2008) A trust and reputation model for filtering and classifying knowledge about urban growth. GeoJournal 72(3–4):229–237

Budhathoki NR, Bruce B (Chip), Nedovic-Budic Z (2008) Reconceptualizing the role of the user of spatial data infrastructure. GeoJournal 72(3):149–160. Available at: http://link.springer.com/article/10.1007%2Fs10708-008-9189-x?LI=true#page-2

Burns CM, Vicente KJ (1996) Judgements about the value and cost of human factors information in design. Inf Process Manage 32(3):259–271

Cooper AK (2011) Challenges for quality in volunteered geographical information. In: AfricaGEO 2011. AfricaGEO, Cape Town, p 13

Coote A, Rackham L (2008) Neogeography data quality—is it an issue? In: Holcroft C (ed) Proceedings of AGI geocommunity'08. Association for Geographic Information (AGI), Stratford-Upon-Avon, p 1. http://www.agi.org.uk/SITE/UPLOAD/DOCUMENT/Events/AGI2008/Papers/AndyCoote.pdf

Crone GR (1968) Maps and their Makers: an introduction to the history of cartography, 4th edn. W. G. East (ed), Hutchinson, London

Das T, Kraak MJ (2011) Does neogeography need designed maps? In: Proceedings of the 25th international cartographic conference and the 15th general assembly of the international cartographic association. International Cartographic Association (ICA), Paris, pp 1–6. Available at: http://www.itc.nl/Pub/GIP/GIP-Academic-Output/GIP-Output.html?l=6&y=2011&d=GIP

DeGarmo EP, Black JT, Kohser RA (2003) Materials and processes in manufacturing, 9th edn. Wiley, New York

Doan A, Ramakrishnan R, Halevy AY (2011) Crowdsourcing systems on the world-wide web. Commun ACM 54(4):86–96. Available at: http://cacm.acm.org/magazines/2011/4/106563-crowdsourcing-systems-on-the-world-wide-web/fulltext

Flach JM (1998) An ecological approach to interface design. In: Proceedings of the 42nd annual meeting of the Human Factors and Ergonomics Society, San Antonio, pp 295–299

Flanagin AJ, Metzger MJ (2007) The role of site features, user attributes, and information verification behaviors on the perceived credibility of web-based information. New Media Soc 9(2):329–342. Available at: http://nms.sagepub.com/content/9/2/319.abstract

Flanagin AJ, Metzger MJ (2008) The credibility of volunteered geographic information. GeoJournal 72:137–148

Goodchild MF (2007a) Citizens as sensors: the world of volunteered geography. GeoJournal 69(4):211–221. Available at: http://www.springerlink.com/content/h013jk125081j628/

Goodchild MF (2007b) Citizens as voluntary sensors: spatial data infrastructure in the world of Web 2.0. Int J Spatial Data Infrastruct Res 2:24–32

Goodchild MF (2008) Commentary: whither VGI? GeoJournal 72(3):239–244

Goodchild MF, Glennon JA (2010) Crowdsourcing geographic information for disaster response: a research frontier. Int J Digit Earth 3(3):231–241

Grira J, Bédard Y, Roche S (2010) Spatial data uncertainty in the VGI world: going from consumer to producer. Geomatica 64(1):61–72

Haklay M (2010) How good is volunteered geographical information? a comparative study of openstreetmap and ordnance survey datasets. Env plann B 37(4):682–703

Haklay M, Ather A, Basiouka S (2010a) How many volunteers does it take to map an area well? In: Haklay M, Morley J, Rahemtulla H (eds) Proceedings of the GIS research UK 18th annual conference. University College, London, p 193–196

Holone H, Misund G, Holmstedt H (2007) Users are doing it for themselves: pedestrian navigation with user generated content. In: Al-Begain K (ed) NGMAST 2007: international conference on next generation mobile applications, services and technologies. IEEE, Cardiff, pp 91–99. Available at: http://ieeexplore.ieee.org/stamp/stamp.jsp?tp=&arnumber=4343406

Idris NH, Jackson MJ, Abrahart RJ (2011a) Colour coded traffic light labeling: a visual quality indicator to communicate credibility in map mash-up applications. In: Presented at the international conference on humanities, social sciences, science & technology (ICHSST). Manchester, pp 1–7. Available at: http://www.geoinfo.utm.my/geoinformatic/geoinformaticpublications/2011/Nurul_Idris_ICHSST_new.pdf

ISO 9241-11 (1998) Ergonomic requirements for office work with visual display terminals (VDT)s—Part 11: guidance on usability. International Standards Organisation, Geneva

Keen A (2007) The cult of the amateur. Nicholas Brealey, Finland

References

Koops MA (2004) Reliability and the value of information. Anim Behav 67(1):103–111. Available at: http://www.sciencedirect.com/science/article/pii/S0003347203004123

Metzger MJ, Flanagin AJ (2011) Using Web 2.0 technologies to enhance evidence-based medical information. J Health Commun 16(Suppliment 1):45–58

Mihalcea R, Chklovski T (2003) Building sense tagged corpora with volunteer contributions over the web. In: Nicolov N et al (eds) Recent advances in natural language proceedings III. John Benjamins, Philadelphia, pp 357–367

Monmonier M (2006) From squaw tit to whorehouse meadow. The University of Chicago Press, USA

Mummidi L, Krumm J (2008) Discovering points of interest from users' map annotations. GeoJournal 72:215–227

Norman DA (2005) Emotional design. Basic Books, USA

Obermeyer N (2007) Thoughts on volunteered (Geo)slavery. In: Goodchild MF, Gupta R (eds) NCGIA and Vespucci workshop on volunteered geographic information. The National Center for Geographic Information and Analysis, Santa Barbara, pp 13–14. http://www.ncgia.ucsb.edu/projects/vgi/docs/position/Obermeyer_Paper.pdf

Ordnance Survey (2009) New revision programme for large-scale topographic data. Ordnancesurvey.co.uk, 2009(Nov 3rd). Available at http://www.ordnancesurvey.co.uk/oswebsite/products/osmastermap/information/technical/revisiontopographic.html. Accessed 3 Nov 2009

Petty, Richard E, Cacioppo, John T (1986) "The Elaboration Likelihood Model of Persuasion." Advances in Experimental Social Psychology, 19: 123–162

Preece J (2002) Interaction design: beyond human-computer interaction. Wiley, New York

Pultar E, Raubal M, Goodchild MF (2008) GEDMWA: geospatial exploratory data. In: Proceedings of the 16th ACM SIGSPATIAL international conference on advances in geographic information systems (ACM GIS 2008). ACM, Irvine, p 499

Ray NM, Ryder ME (2003) "Ebilities" tourism: an exploratory discussion of the travel needs and motivation of the mobility-disabled. Tour Manag 24:57–72

Reeves S, Sherwood S (2010) Five design challenges for human computation. In: Proceedings of the 6th nordic conference on human-computer interaction: extending boundaries. ACM, Iceland, p 383

Rieh SY (2002) Judgment of information quality and cognitive authority in the Web. J Am Soc Inf Sci Technol 53(2):145–161. Available at: http://onlinelibrary.wiley.com/doi/10.1002/asi.10017/full

Rogers EM (2003) Diffusion of innovations, 5th edn. Free Press, New York

Schilder F, Habel C (2001) From temporal expressions to temporal information: Semantic tagging of news messages. In: Proceedings of the workshop on temporal and spatial information processing, vol 13. Association for Computational Linguistics, pp 1–8

Sheridan TB (1995) Reflections on information and information value. IEEE Trans Syst Man Cybern 25(1):194–196

Shirky C (2009) How cellphones, Twitter, Facebook can make history. In: Frawley Bagley E (ed) TED@State. Washington, TED Talks. Available at: http://www.ted.com/talks/clay_shirky_how_cellphones_twitter_facebook_can_make_history.html

SozialHelden (2012) Wheelmap: Loughborough. Wheelmap.org. Available at: http://wheelmap.org/en?zoom=18&lat=52.77183&lon=-1.20694&layers=BT. Accessed 12 Nov 2012

Stephens DW (1989) Variance and the value of information. Am Nat 134(1):128–140

Tapscott D, Williams AD (2008) Wikinomics: How Mass Collaboration Changes Everything. Atlantic Books, UK

Tsou M-H (2005) An intelligent software agent architecture for distributed cartographic knowledge bases and internet mapping services. In Peterson MM (ed) Maps and the internet. Elsevier Ltd., Oxford, pp 229–243

Van Exel M, Dias E (2011) Towards a methodology for trust stratification in VGI. In: VGI preconference at AAG. Association of American Geographers, Seattle, pp 1–4

Warnick B (2004) Online ethos: source credibility in an authorless environment. Am Behav Sci 48(2):256–265

Wilson P (1983) Second-hand knowledge: an inquiry into cognitive authority. Greenwood Press, Connecticut

Wood D (2003) Cartography is dead (thank god!). Cartograph Perspect 45(Spring):4–7. Available at: http://makingmaps.owu.edu/mm/cartographydead.pdf

Zook MA (2010) Volunteered geographic information and crowdsourcing disaster relief: a case study of the Haitian earthquake. World Med Health Policy 2(2):7–33

Further Reading

Brown M, Sharples S, Harding J, Parker CJ, Bearman N, Maguire M, Forrest D, Haklay M, Jackson M (2012) Usability of geographic information; current challenges and future directions. Appl Ergon 44 (6):855–865

Harding J, Sharples S, Haklay M, Burnett G, Dadashi Y, Forrest D, Maguire M, Parker CJ, Ratcliff L (2009) Usable geographic information—what does it mean to users?. In: Proceedings of the AGI GeoCommunity '09 conference, Stratford-Upon-Avon, UK, AGI GeoCommunity

Parker CJ (2012) A human factors perspective on volunteered geographic information. Loughborough University, UK

Parker CJ, May AJ, Mitchell V (2010) An exploration of volunteered geographic information stakeholders. In: Haklay M, Morley J, Rahemtulla H (eds) Proceedings of the GIS research UK 18th annual conference, University College London: UCL, pp 137–142

Parker CJ, May AJ, Mitchell V (2011) Relevance of volunteered geographic information in a real world context. In: Proceedings of GIS research UK 19th annual conference, University of Portsmouth: GIS Research UK, pp 230–236

Parker CJ, May AJ, Mitchell V (2012a) The role Of VGI and PGI in supporting outdoor activities. Appl Ergon 44 (6):886–894

Parker CJ, May AJ, Mitchell V (2012b) Understanding design with VGI using an information relevance framework. Trans GIS 16 (4):545–560

Parker CJ, May AJ, Mitchell V (2013) User centred design of neogeography: the impact of volunteered geographic information on trust of online map Mashups. Ergonomics (In Press)

Parker CJ, May, AJ, Mitchell V (2012c) Using VGI to enhance user judgements of quality and authority. In: Whyatt D, Rowlingson B (eds) Proceedings of GIS research UK 20th annual conference, Lancaster: GIS Research UK, pp 171–178

Parker CJ, May AJ, Mitchell V, Burrows A (2013) Capturing volunteered information for inclusive service design: potential benefits and challenges. Des J 16(2):197–218

Index

A
Accessibility, 60
Accuracy, 4, 39, 60
Affectiveness, 60
Availability, 60

B
Base map, 16

C
Clarity, 60
Conditional, 41
Crowdsourcing, 122
Currency, 60, 115

D
Data content, 121
Data generation, 79
Data richness, 5
Depth, 60
Design recommendations, 130
Digital earth, 4

E
Emotion, 34, 41
Epistemic, 41
Experiences, 74
External information, 60

F
Focus groups, 51
Framework of VGI, 125
Function, 35
Functional, 41

G
Geocollaboration, 5.
    See Neogeography
Geographic information, 3
Geoweb. *See* Neogeography
GIS, 16
GPS, 1

H
Hierarchical task analysis, 56

I
Impact, 95
Influences of VGI, 127
Information access, 72
Information currency, 70
Information judgements, 124
Information use, 39, 49
Inter-user data flow, 33

J
Judgements, 97

K
Knowledge, 36
Knowledge of VGI presence, 114

L
Legal, 37, 41
Limitations of VGI, 128
Locative media. *See* Neogeography

M
Map hacking. *See* Neogeography
Mapping party, 28
Mashups, 87, 96. *See* Neogeography
Missing data, 6
Mobile computing, 123
Moral, 37, 41
Motivation, 122

N
Neogeography, 2, 11

O
Objectivity, 18, 126
Openstreetmap, 15, 27

P
Participatory observation, 51
PGI, 12, 16
Price, 37, 41
Production of VGI, 122
Public participatory GIS. *See* Neogeography

Q
Quality, 60
Quality and authority, 98, 113
Quality control, 90, 126

R
Real time information, 71
Relevance, 51
Rich picture, 43

S
Smartphone, 122
Social, 38, 41
Soderini letter, 1
Spatial crowdsourcing. *See* Neogeography
Spatial media. *See* Neogeography

T
Tangibility, 60
Tracks4Africa, 16
Trust, 6, 49, 72

U
Uniqueness, 90
Usability, 103
User groups, 30
User interaction, 43
User perceptions, 95
User relationships, 33

V
Value, 13, 30, 34
Verification, 60
VGI, 2, 12, 16
Volunteered information, 4

W
Web 2.0, 1, 4
Wikipedia, 16, 113

Printed in Germany
by Amazon Distribution
GmbH, Leipzig